Microbes That Shaped Our World

# Microbes That Shaped Our World

What We Must Know About Living With Microbes

*By*

Gustaaf Hallegraeff

DE GRUYTER

ISBN 978-3-11-914233-5
ISBN 978-3-11-222826-5 (PDF)
ISBN 978-3-11-222827-2 (E-PUB)
DOI https://doi.org/10.1515/9783112228265

Library of Congress Control Number: 2026947740

Bibliographic information published by the Deutsche Nationalbibliothek
The Deutsche Nationalbibliothek lists this publication in the Deutsche Nationalbibliografie; detailed bibliographic data are available on the Internet at https://dnb.dnb.de.



De Gruyter and Walter de Gruyter GmbH are part of De Gruyter Brill.
www.degruyterbrill.com

Questions about General Product Safety Regulation: productsafety@degruyterbrill.com

Cover illustration: Gustaaf Hallegraeff

# Contents

About the Author IX
Acknowledgements XI
List of Figures XIII

1 Introduction 1

2 What are Microbes? 5

3 Kingdoms of Life 9

PART 1
Cyanobacteria

4 Lungs of the Planet: First Cyanobacteria 13

5 *Prochlorococcus:* Ocean's Smallest Photosynthetic Cell 17

6 *Trichodesmium* "sea sawdust" 19

7 Cyanobacterial Blooms and Drinking Water 21

8 *Spirulina* and the Aztecs 25

PART 2
Algae

9 Microscopic Algae: Grasses of the Sea 29

10 Coccolithophorid *Emiliania huxleyi* and Gaia 33

11 *Phaeocystis* and Sea Foam 37

12 *Phytophthora infestans* and Irish Potato Famine 39

13 Diatoms: Miniature Architects 41

14 Blue Diatom *Haslea ostrearia:* Oysters Fit for Royalty 45

15 *Didymosphaenia geminata* or “rock snot” 47

16 *Pseudo-nitzschia* and Hitchcock’s *The Birds* 49

17 Diatom Ooze and Alfred Nobel’s Dynamite 51

18 Dinoflagellates: Masters of Chemical Warfare 53

19 *Alexandrium* and Captain Vancouver’s Poison Cove 55

20 *Karenia brevis* and Florida Red Tides 57

21 *Gambierdiscus* and Ciguatera Poisoning 59

22 *Noctiluca* Night Lights 61

23 Harmful Algal Blooms and Finfish Kills 63

24 Zooxanthellae and Coral Bleaching 65

25 Green Algae: The Origin of Land Plants 67

26 *Dunaliella salina* and β-carotene 69

27 *Botryococcus*: Good and Bad Oils 71

PART 3
Protozoa

28 Radiolarians: Art Forms in Nature 75

29 Tintinnid Loricas: Master Home Decorators 79

30 *Plasmodium* and Malaria 81

PART 4
Fungi

Introduction to Part 4 85

31 Baker's Yeast *Saccharomyces* 87

32 *Penicillium*: The Birth of Antibiotics 89

33 Soil Fungi and the Wood Wide Web 91

34 Fungal Toxins in Food 93

35 Fungal Pathogens Impacting Viticulture 95

36 *Aspergillus*, Coral Disease and Dust Storms 97

PART 5
Bacteria

Introduction to Part 5 101

37 Human Microbiome: You Are What You Eat 103

38 *Yersinia pestis* and the Black Death 105

39 *Vibrio* and Cholera 107

40 *Bacillus anthracis* and Biological Warfare 109

41 *Mycobacterium* and Tuberculosis 111

42 *Legionella pneumophila* and Legionnaire's Disease 113

43 Soil Bacteria 115

PART 6
Viruses

Introduction to Part 6 119

44 *Variola*, Smallpox and Monkeypox 121

45 *Morbillivirus* and Measles 123

46 *Influenza* and the Spanish Flu 125

47 *Zika* Virus 127

48 *Pilchard Herpes Virus* (PHV) 129

49 *Ostreid Herpes Virus* and Pacific Oyster Mortality 131

50 *Apthovirus* and Foot and Mouth Disease in Cattle 133

51 *Leporipoxvirus* and Myxomatosis 135

52 Bacteriophages 137

53 *Human Immunodeficiency Virus* and AIDS 139

54 Coronaviruses and COVID-19 141

PART 7
What Can We Learn from Microbes?

55 The Microbial World Darwin Never Knew 145

56 Microevolution: The Power of Growing Fast 147

57 Microbes and their Toxins 149

58 Quarantine for Ships, Aquaculture and Agriculture 151

59 Quarantine for Human Disease Outbreaks 153

60 Microbes and Climate Change 155

Epilogue 157
Glossary 159
Bibliography 167
Index of Subject 181
Index of Species 185

## About the Author

Gustaaf Hallegraeff grew up in a Dutch country village, some 30 km from Delft, where in 1673 Anton Van Leeuwenhoek using crude glass lenses for the first time observed microscopic "*animalcules*" in pond water. A childhood gift of a toy microscope led to his own life changing moment, peering at the whirling green creatures in a backyard rainwater tank. Neither his parents nor schoolmaster were able to explain whether these were plants or animals or if they were benign or dangerous. He was hooked for life, and Van Leeuwenhoek became his all-time hero. He completed a Master's degree at the University of Amsterdam working on diatoms from krill recovered from Antarctic whales, and a Doctoral degree on microscopic algal plankton of Dutch freshwater lakes. The lure of the blue ocean and a sunnier climate motivated his move to a job at the Australian Commonwealth Fisheries and Oceanography laboratories in Sydney. In 1998 he became a Professor at the University of Tasmania, and head of the School of Plant Science, and later a founding member of the Institute for Marine and Antarctic Studies (IMAS), where he supervised 40 PhD students working on microscopic algae. During the 1990s, scientific appreciation of the global importance of microscopic plankton escalated. Intensified research on climate change caused us to revise estimates of the contribution of microalgae to global photosynthesis from the 25% mooted in his old textbooks to "every second breath of oxygen we inhale comes from ocean plankton". The advent of molecular genetics in the 2000 s made us aware that the genetic diversity of microscopic organisms far exceeds that of land plants and animals. Most of his career focused on harmful algal blooms that damage tourism, threaten aquaculture operations and can poison human consumers of seafood products. The 2019 COVID experience motivated him to expand his interest also into the role of fungi, bacteria and viruses in shaping the microbial engine of life on our unique planet.

 | HTTPS://DOI.ORG/10.1515/9783112228265-203

The author kneeling on the grave of Van Leeuwenhoek inside the Oude Kerk (Old Church) in Delft, Netherlands. The inscription on the grave stone translates as: "Here rests Anthony van Leeuwenhoek, oldest member of the Royal Society in London. Born in the city of Delft on 24th of October 1632 and died on the 26th of August 1723 at the age of 90 years, 10 months and 2 days. To the reader: That all visitors may respect old age and miraculous talents, so thread respectfully. Here lies buried the scientific wisdom of Leeuwenhoek".

# Acknowledgements

I acknowledge my late uncle, veterinarian Peter Beersma for the childhood gift of my first toy microscope, and Professors Siebrecht van der Spoel and the late Joop Ringelberg at the University of Amsterdam for teaching me ocean plankton biogeography and freshwater plankton ecology, respectively. The late Shirley Jeffrey at CSIRO Fisheries & Oceanography in Australia introduced me to the joy of culturing microscopic algae. Above all, I thank my many students at the University of Tasmania, for working with me and expanding my interests into the areas of molecular genetics (Dr Miquel de Salas, Dr Linda Armbrecht), marine bacteria (A/Prof Chris Bolch), fungi (Dr Aiko Hayashi) and viruses (Prof Andrew McMinn). I thank my partner Cawley Farrell and daughter Isabella for their patience and understanding when at times I appeared to be living in a different world. All illustrations are by the author in collaboration with his students and collaborators, unless otherwise indicated.

Author Photograph: Steven Hallegraeff; Fig. 1. Oil painting of Leeuwenhoek by J. Verkolje 1686. Reproduced from an original held at Museum Boerhaave, Leiden, Netherlands; Public Domain; Leeuwenhoek's drawings from *Philosophical Transactions Royal Society London* 283, 1304–1311 (1703). CC-BY-4.0; Fig. 4. Toxic bloom warning poster from New York State Department of Health; Fig. 6. White cliffs of Dover: courtesy Chris Rees; Fig. 7. Sea foam: Steven Hallegraeff; *Phaeocystis* light micrograph by National Institute for Environmental Studies, microbial culture collection, Japan; Fig. 10. Hitchcock's movie "The Birds": Universal Pictures theatrical poster 1963; Fig. 13. Toxic shellfish poster by the Tasmanian Government Department of Health and Human Services; Fig. 14. Ciguatera fish poisoning poster by the Florida Department of Health; Fig. 15. Bioluminescence in Southern Tasmania: courtesy James Garlick; Fig. 16. *Acropora* staghorn coral; image by Behan, via Flickr; Microscopic zooxanthella: image Todd LaJeunesse, Oregon State University, Wikipedia, CC-BY-SA-2.0; Fig. 17. Ciliate *Paramecium bursarium*: image Anatoly Mikhaltsov, Wikipedia, CC-BY-SA-4.0; Fig. 18. Betatene pond culture of the green alga *Dunaliella salina* in Western Australia. Pond cultures of *Spirulina*: Livefuel, California; Fig. 19. Buckminster Fuller's geodesic dome: image by Alamy; René Binet's Porte Monumentale in Paris: Léopold Mercier / Roger-Viollet (CC BY-SA); Fig. 21. Scanning electron micrographs of tintinnid loricae, *Codonella galea*: image Karine Leblanc, Aix-Marseille University, France; *Codonellopsis pusilla*: image Ruth Eriksen, CSIRO, Australia; *Dictyocysta lepida*: image Ruth Eriksen, CSIRO, Australia; Fig. 22. Mosquito *Aedes aegypti*: image iStock frank600; *Plasmodium*: image iStock Christoph Burgstedt; Fig. 23. Yeast: Science Photo Library/Steve Gschmeissner/Getty Images; Fig. 24. Commercial penicillin: Advanced Allergy Asthma Clinic, https://www.allergyasthma.clinic. *Penicillium* fungal spores: image: AJ Cann | Flickr;

 | HTTPS://DOI.ORG/10.1515/9783112228265-204

Fig. 25. Forest floor fungal mycelium: https://james-camerons-avatar.fandom.com; *Rhizobium* in root nodules: image Alamy; Fig. 26. Dubrovnik quarantine of ships: www. theravenreport.com; 27 March 2017; Bacterium *Yersinia pestis*: image iStock Dr Microbe; Fig. 27. Arthur Hassall's *Microscopic Examination of the Water Supplied to the Inhabitants of London*, 1850: Wellcome Library, London; *Vibrio cholera* bacteria: Alamy Stock Photo; Plate 28. Shutterstock images of chimpanzee, bat, rat, flee; Aids virus: by Hans R. Gelderblom, Robert Loch Institut, Germany; Fig. 29. Face masks: Asashi Shimbun file photo; https://www.asahi.com/ajw/articles/14302356); Corona virus: Alamy Stock Photo; Fig. 30. 1987 Grim Reaper TV commercial: Grey Advertising; screen shot; AIDS virus: https://africacdc.org/disease/hiv-human-immunodeficiency-virus/; AIDS Poster 1987; London School of Hygiene & Tropical Medicine collection.

# Figures

Author Photograph The author kneeling on the grave of Van Leeuwenhoek inside the Oude Kerk (Old Church) in Delft, Netherlands. The inscription on the grave stone translates as: "Here rests Anthony van Leeuwenhoek, oldest member of the Royal Society in London. Born in the city of Delft on 24th of October 1632 and died on the 26th of August 1723 at the age of 90 years, 10 months and 2 days. To the reader: That all visitors may respect old age and miraculous talents, so thread respectfully. Here lies buried the scientific wisdom of Leeuwenhoek". x

1 Top left: Oil painting of Van Leeuwenhoek by J. Verkolje 1686; Top right: Detail of replica of one of Van Leeuwenhoek's stamp size microscopes; Bottom left: Van Leeuwenhoek's drawings of (left) diatoms, ciliates, rotifers and hydroids attached to duckweed from a Delft canal; From *Philosophical Transactions Royal Society London* 283, 1304-1311 (1703). Bottom middle: Bacteria from his own mouth with the dotted line in fig.B portraying movement; Bottom right: Sperm cells. 7

2 Left. Haeckel's (1866) early tree of life positioned the higher plants and higher animals at the crown of evolution but for the first time also recognised the existence of a third Kingdom, the Protista. A radically different view of the phylogenetic tree of life in the 2000s, based on molecular sequences. Shown are the three domains Archaea, Bacteria (including the sulphur bacteria from which originated the mitochondria, and the cyanobacteria from which originated the chloroplasts) and the Eukaryota (with a true nucleus). The latter include the Kingdom of Protists such as diatoms, dinoflagellates and ciliates, Kingdom of Fungi, Kingdom of Plants and Kingdom of Animals. In terms of genetic diversity, life on our planet is dominated by microbes, with plants and animals and humans representing very small branches only. 10

3 Top: Banded iron formations reflecting the Great Oxidation Event, resulting from the advent 3.5 billion years ago of cyanobacterial photosynthesis. Original photograph from specimen held at Artis Groote Museum in Amsterdam. Bottom: Scanning micrograph of extant filamentous cyanobacteria *Nodularia spumigena* from an Australian estuary. 15

4 Cyanobacterial bloom of *Microcystis aeruginosa* from a Dutch canal. Inset: Toxic bloom warning poster from New York State Department of Health. 23

5 Top. NASA Satellite image of surface chlorophyll-a phytoplankton pigments during the Northern Hemisphere spring diatom blooms (highest concentrations in yellow and orange). Below: Light micrograph of diatom spring bloom in South Australian waters dominated by chain-forming colonies of *Skeletonema* and *Chaetoceros*. 31

6 Top: White cliffs of Dover comprised of the calcareous remains of microscopic planktonic coccolithophorids. Bottom: Scanning electron micrograph of the cultured coccolithophorid *Emiliania huxleyi*. 35

 | HTTPS://DOI.ORG/10.1515/9783112228265-205

7 Top: Sea foam on the beaches in the Netherlands caused by *Phaeocystis* golden-brown flagellate blooms. Bottom: Light micrograph of cultured cells. 38
8 Diatoms. Plate 84 from Haeckel, E. 1899-1904. *Künstformen der Natur.* 43
9 Top: Brown "rock snot" from the Buller River, New Zealand, caused by the growths of the diatom *Didymosphaenia geminata.* Bottom: Scanning electron micrograph of the large bottle-shaped diatom cells. 48
10 Hitchcock's movie "The Birds" was based on an incident in California in 1961 when crazed seabirds pelted homes, unable to fly and regurgitating anchovies, which had been contaminated with toxins from a *Pseudo-nitzschia* diatom bloom. 50
11 Scanning electron micrograph of diatom rich sediments ("diatomaceous ooze") from Adelie Shelf, Antarctica. 52
12 Dinoflagellates. Plate 14 from Haeckel, E. 1899-1904. *Künstformen der Natur.* 54
13 Toxic shellfish warning poster by the Tasmanian Government Department of Health and Human Services, against the background of a red tide by the dinoflagellate *Noctiluca scintillans* threatening to suffocate circular salmon farming pens in Southern Tasmania. 56
14 Ciguatera fish poisoning information poster by the Florida Department of Health, against the background of a scanning electron micrograph of a *Gambierdiscus lapillus* dinoflagellate culture from the Australian Great Barrier Reef. 60
15 Top: Bioluminescence spectacle in Southern Tasmania. Bottom: Light micrograph of a Tasmanian bloom sample of the causative dinoflagellate *Noctiluca scintillans.* 62
16 Top: *Acropora* staghorn coral in the process of bleaching. Coral polyps on the outer branches have expelled their symbiotic algae, losing a source of food as well as their colour. Bottom: Microscopic image of symbiotic zooxanthella dinoflagellates responsible for the entire productivity of coral reef ecosystems. 66
17 Top: The protozoan ciliate *Paramecium bursarium* which feeds on coccoid green *Chlorella* type algae but which it keeps alive as in a green house. This is how the green chloroplasts of land plants originated. Bottom: Light micrographs of two species of the colonial freshwater green alga *Cosmarium.* 68
18 Examples of algal biotechnology. Top. Betatene pond culture of the green alga *Dunaliella salina* in Western Australia used for the production of ß-carotene. Bottom. Pond cultures of the cyanobacterium *Spirulina* for biofuel production in the Californian desert. 72
19 Architects using protists for inspiration. Top: Buckminster Fuller's geodesic dome structure as used for the Montreal 1967 Expo building bears a striking resemblance to the diatom cell wall of *Pyxidicula* (resting spore of *Stephanopyxis*). Middle: The leaning tower of Pisa resembles the chain forming diatom *Paralia sulcata*; Bottom: René Binet's Porte Monumentale at the Place de la Concorde in Paris which served as the main entrance to the 1900 world Expo. This portal was modelled on radiolarian *Clathrocanium reginae.* (see Fig.20). 76

20 Radiolaria. Plate 31 from Haeckel, E. 1899-1904. *Künstformen der Natur.* 77

21 Scanning electron micrographs of planktonic tintinnid loricae decorated with coccolithophorids from the surrounding seawater. Top left: *Codonella galea* from tropical Tonga waters; Top right: *Codonellopsis pusilla* from the Southern Ocean; Bottom left: *Codonellopsis pura* from Coral Sea; Bottom right: *Dictyocysta lepida* from Southern Ocean. 80

22 Top: Mosquito *Aedes aegypti* sucking human blood. Bottom: *Plasmodium* protozoan parasite (green) among human red blood cells. 82

23 Yeast is one of the earliest domesticated microbes, dating back 5,000 yrs ago. Without yeast fungus, we would not be able to enjoy bread or wine or beer. Without *Penicillium* fungus we would not have French cheese. 88

24 Commercial penicillin product. Below: Scanning micrograph of *Penicillium* fungal spores. 90

25 Top left. The forest floor microbiome comprises a fungal mycelium as visualised in the movie *Avatar* "Tree of Souls". Top right: nitrogen fixing bacteria such as *Rhizobium* in root nodules. Bottom: Fungal mycelium and fungal spores of *Aspergillus sydowii* from an Australian dust storm event. 92

26 The concept of "quarantine" whereby ships were forced to wait offshore for 40 days before entering the city of Dubrovnik, then under Venetian rule, dates back to 1348 when the Great Plague was devastating Europe. Bottom: The bacterium *Yersinia pestis*, causative organism of the Black Plague. 106

27 Top: Illustrations from Arthur Hassall's *Microscopic Examination of the Water Supplied to the Inhabitants of London*, 1850. Bottom: Scanning micrograph of *Vibrio cholera* bacteria. 108

28 When humans encroach on animal habitats that harbour viruses and bacteria, this can cause microbes to jump species barriers and infect humans, e.g. AIDS virus originated from chimpansees and COVID virus most likely from bats. Flees and rats can play a role in transfer of microbes to humans, notably with *Yersinia pestis* bacteria responsible for the Plague, and Malaria *Plasmodium* protists transferred by mosquitoes. 140

29 The COVID-19 pandemic made us familiar with face masks, Rapid Antigen Tests, and the Corona Virus. 142

30 Most human disease pandemics start with incomplete knowledge and fear leading to misinformation and scare campaigns. The 1987 Grim Reaper TV commercials in Australia demonised the broader LGBTQ+ community, but today with major scientific advances in testing, prevention and treatment, living with HIV is no longer a death sentence and our outlook on this problem has changed remarkably. 146

CHAPTER 1

# Introduction

We humans are well familiar with how plants such as corn, grape, potato, rice, tomato and tulip shaped our world [1]. Similarly, we benefited immensely from our association with animals such as bees, cats, cod, cows, chickens, dogs, herring, horses, oysters, sheep and whales [2]. However, the role that invisible microbes play in shaping our world, both good – the oxygen we breathe, the seafood we eat, baker's yeast, the source of antibiotics, the human gut microbiome – and bad – harmful algal blooms, bacterial and viral pandemics – is not widely appreciated.

Until Van Leeuwenhoek's first microscopic observations in 1673, humans were completely unaware of the existence of microbes. Using crude glass lenses, this Dutch clothing merchant newly became aware of the existence of little animals (*animalcules*) in Dutch pond water, even in his mouth, his stool and in his own sperm. German scientist Ernest Haeckel in 1866 first proposed that these smallest creatures did not fit into the Kingdoms of Plants nor Animals. To accommodate them he coined the Kingdoms of Protista (protozoa, algae, slime moulds) as well as Monera (bacteria). In 1969 Robert Whittaker from Cornell University further separated the Kingdom Fungi from the Plantae, and in 1977 Carl Woese from the University of Illinois, based on newly available molecular sequences, divided the Bacteria into Archaea and Eubacteria (the latter including the blue-green algae or cyanobacteria). It was another Dutchman, Martinus Beyerinck, working in 1897 on tobacco disease, who when unable to detect a bacterial origin first postulated the involvement of something even smaller, now known as Tobacco Mosaic Virus. Viruses became not visible until 1931 with the first use of the electron microscope. Even though containing molecular DNA or RNA, viruses are often excluded from the tree of life because they must hijack host cells before they can reproduce or synthesize proteins.

Early microbiologists espoused the wonders and beauty of microbes, such as Haeckel's magnificently illustrated "*Artforms in Nature*" [3,4,5]. Sadly, over time, humans have grown increasingly distant from nature and instead tend to focus on the negative impacts of microbes. On average only one in about 1,000 microbes are harmful to humans. All the 100 or more microbes covered in this book predated the arrival of *Homo sapiens* on this planet by billions or millions of years. Any harm they caused us humans – algal toxins contaminating seafood, cyanobacteria contaminating drinking water, bacterial and viral diseases such as smallpox, plague, COVID – have been purely accidental. Microbes never meant to kill us. There are no benefits of doing so. Pathogenic viruses need a host to infect, reproduce and disperse. Ebola is not a clever virus because it kills its host too quickly, in contrast to AIDS and COVID which piggy-backed on an ever growing human population and

 | HTTPS://DOI.ORG/10.1515/9783112228265-001

readily spread via increased global travel. The repercussions of "germs" for evolution of human civilization have been previously explored by William McNeill's 1976 *Plagues and People,* Jared Diamond's 1997 *Guns, Germs, and Steel,* and reiterated by Jonathan Kennedy's 2023 *Pathogenesis* [6,7,8]. The Black Death, by the bacterium *Yersinia pestis,* curtailed the expansion of the Roman Empire. The 1918 Influenza pandemic killed 50 million people, more than during the entire World War 1. AIDS signaled the end of the 1960s free love era, while in 2019 COVID taught us to work from home by Zoom. Microbes have shaped not only our planet but our human civilization. Plagues have been known since historical times such as described in the Christian Bible *Book of Exodus* 1000 BC ("*sores that break into pustules on man and beast*"; "*all the waters that were in the river were turned to blood. And the fish that were in the river died*"). In the absence of any scientific understanding of their causes, they were often viewed as punishments by God, but such fear is no longer justified. Imagine having to cope with the COVID pandemic at a time when microscopic life was completely unknown. With the advent of rapid molecular techniques and increasing computational capacity, we now can speedily identify the causative organisms, understand who they are and what they are doing, and adjust human behavior to mitigate impacts. Three quarters of emerging human diseases have confirmed origins in animals, either wildlife – bats, primates, rodents and birds – or domesticated animals – poultry, pigs and camels. Measles originated from rinderpest, smallpox is related to cowpox, AIDS originated from African primates, COVID from bats, and so on. Climate change enhances the spread of diseases transmitted by mosquitoes (e. g., malaria, yellow fever). Increased ease of global plane travel makes spread of airborne diseases more likely than ever. Changes in land use, deforestation, agricultural expansion, trade in wildlife and ever expanding human settlements are escalating the impacts on humans.

This book aims to strike a balance by equally emphasizing numerous positive applications of learning from microbes to benefit human society. In 1928 Alexander Fleming returned from holiday and noticed how contamination by the fungus *Penicillium* prevented his golden staph bacterial culture plates from growing. The purification of the causative chemical substance penicillin, spearheaded by University of Oxford's Howard Florey, is one of the success stories in the history of medicine. In 1796 English physician Edward Jenner noticed that milk maids seldom caught smallpox. He correctly interpreted this to be due to regular contact with the milder symptoms of cowpox which conferred some form of immunity. This insight led to the development of vaccines (*vacca* is Latin for cow). Diseases such as malaria have driven evolution of the human genome, and identification of the molecular basis of malaria disease resistance in native sub-Saharan human populations is newly being explored toward a new vaccine under trial. In 1485 in Venice the plague bacterium *Yersinia pestis* was responsible for the creation of the strategy of "quarantine" whereby ships, people and goods were held in isolation for 40 days (*quaranta*). Quarantine practices are now the cornerstone of organized responses to infectious disease outbreaks. Early ad hoc counter measures to combat the Spanish flu epidemic in 1918 compare strikingly with those put in place in 2019 for COVID. Social distancing measures were introduced early such as closing schools, theatres and churches, limiting public transportation and banning mass gatherings. Wearing face masks became common in some places such as Japan, but even then there were debates over their efficacy and some resistance to their use. Spread of misinformation and bans on reporting dead tolls were rife. Vaccines against the flu were already attempted in 1918 but they were mistakenly based on bacteria and not the virus, and thus could only help with secondary infections. In 2019 medical science made a big difference by developing vaccines in record time (including the design of novel mRNA vaccines), but on the dark side it created vaccine apartheid between rich and poor countries. Humans are also masters to steal the achievements by microbes for our own military and political advancements. In the 14th century the Mongols catapulted corpses that died of the plague into the city walls of Caffa on the Black Sea. Smallpox contaminated blankets were handed out in 1763 by the British Army to decimate native American Indian populations. Dinoflagellate saxitoxins were used in the 1960s in the United States in CIA suicide capsules, and *Anthrax* bacterial spores were used as biological weapons mailed to US congress men. *Clostridium botulinum* toxin first linked to fatal sausage poisoning in Germany in the 1900s, today is better known from cosmetic Botox anti-wrinkling applications.

At a time when it has become more urgent than ever to rethink how humans cohabit with other forms of life, this book explores microbes as ambassadors to reframe our understanding of the natural world. It is mostly only when our relationships with the natural world are disturbed, that problematic alliances with microbes become apparent. We need microbes' help to solve the human caused climate crisis [9], while realizing they almost certainly will outlive humanity. I hope that you will enjoy the stories of these underrated microscopic creatures and appreciate their illustrations as exquisite art forms which have been sources of inspiration for biologists, conservationists and even architects. Microbes matter! Ignoring them or not taking the effort to fully understand them, we do so at our own peril. This is the story of how microbes shaped our world. This is their story!

CHAPTER 2

# What are Microbes?

Microbes are organisms that are too small to be seen by humans without using a microscope, that means roughly smaller than 1/20 of a millimeter also designated as 50 micrometer (μm) in which 1 micrometer is 1/1,000 mm. We were completely unaware of their existence until 1673. In that year, the Dutch draper and pioneer microscopist Anton Van Leeuwenhoek (Fig. 1), using crude glass lenses, was the first human to see minute creatures in Dutch pond water from Berkelse Mere, which he called 'animalcules'.

> Some of these [animalcules] are so exceedingly small that millions might be contained in a single drop of water. I was much surprised at this wonderful spectacle, having never seen any living creature comparable to these for smallness; nor could I indeed imagine that nature had afforded instances of so exceedingly minute animal proportions.

Van Leeuwenhoek [10] did not develop the first microscope himself, that honor goes to Englishman Robert Hooke [11] with his popular 1665 book *Micrographia: or Some Physiological Descriptions of Miniature Bodies Made by Magnifying Glasses.* While Hooke took common objects and described their appearance up to 30 times magnification, Van Leeuwenhoek perfected the microscope. His hand-held, single-lens microscopes were made by grinding and then polishing glass balls into curved lenses, which allowed him to see objects up to 270 times larger than with the naked eye. His postage stamp size microscopes were simple devices, which had a single lens mounted in a small hole in a brass plate. The specimens were mounted on a sharp needle that stuck up in front of the lens. Leeuwenhoek was a masterly observer who described many types of bacteria, algae and protozoa and calculated their sizes. He also became the first person to observe spermatozoa and one of the earliest to describe red blood cells. He wrote 560 letters to the Royal Society of London, which were translated from Dutch into English and Latin and published in the "Philosophical Transactions of the Royal Society". In 1680, he was elected a Fellow of the Royal Society of England and he was also a corresponding member of the Academy of Sciences in Paris. A public poll held in the Netherlands in 2004 ranked him among the greatest Dutchmen in the same league as Rembrandt, Van Gogh and Vermeer. Fittingly, since 2014 Amsterdam hosts a science museum *Micropia* dedicated to microbes [12] as well as has named a major hospital after Van Leeuwenhoek.

 | HTTPS://DOI.ORG/10.1515/9783112228265-002

However, even the best optical microscopes available today are limited by the wavelengths of visible light, that is, they can magnify no more than 1,000 times and cannot discriminate between objects closer than about 0.2 µm. In 1931, German scientists Ernst Ruska and Max Knoll found a way to achieve a resolution greater than that of light when they realized they could transmit electrons through a specimen to form an image. This led to the first Transmission Electron Microscope (TEM), where electrons are pointed directly at the sample and pass through it to create the image. Ruska employed a similar approach using a focused beam of electrons to scan a sample's surface in a rectangular pattern to deliver information about its topography and composition in which is called a Scanning Electron Microscope (SEM). Scanning microscopes typically achieve magnifications of 20 to 30,000 times, but Transmission Electron Microscopes (TEM) up to 100,000 times. This allowed in the 1930s for the discovery of still smaller microbes, the viruses.

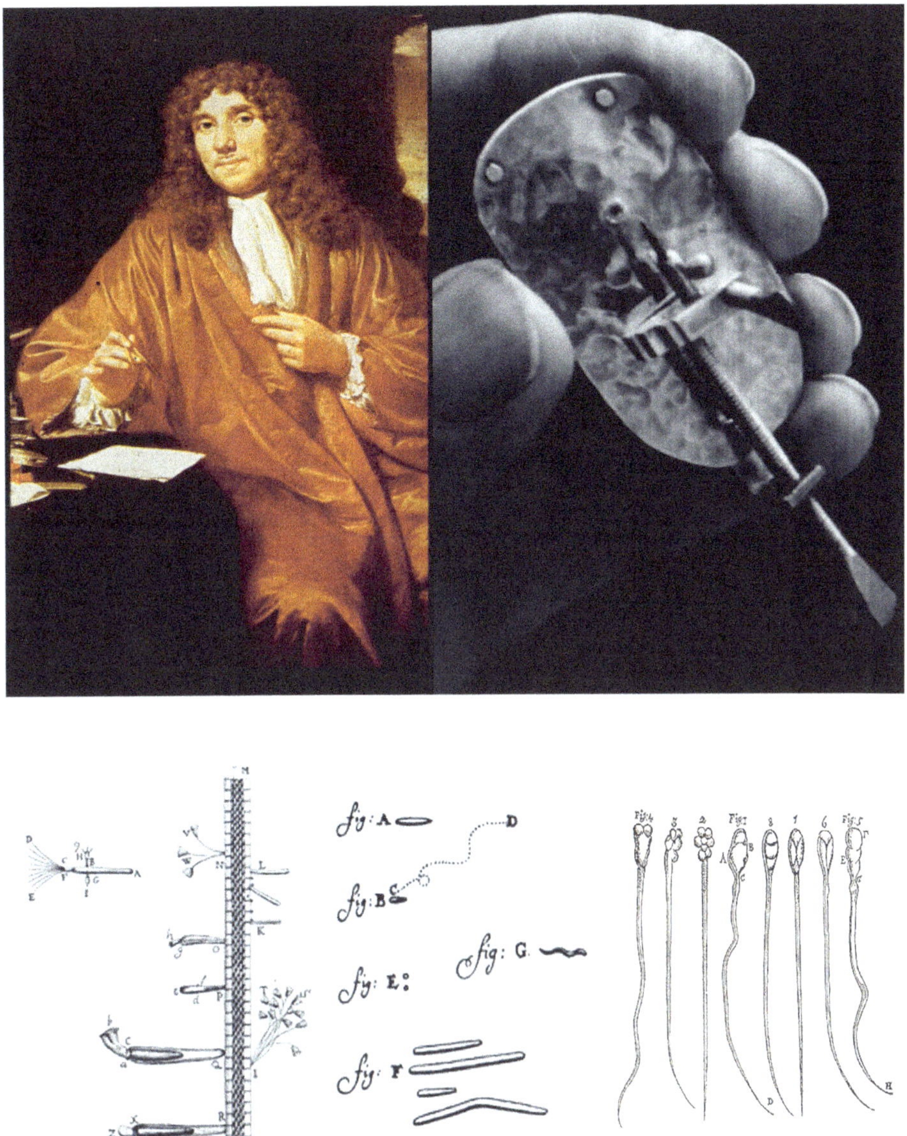


FIGURE 1 Top left: Oil painting of Van Leeuwenhoek by J. Verkolje 1686; Top right: Detail of replica of one of Van Leeuwenhoek's stamp size microscopes; Bottom left: Van Leeuwenhoek's drawings of (left) diatoms, ciliates, rotifers and hydroids attached to duckweed from a Delft canal; From *Philosophical Transactions Royal Society London* 283, 1304–1311 (1703). Bottom middle: Bacteria from his own mouth with the dotted line in fig. B portraying movement; Bottom right: Sperm cells.

CHAPTER 3

# Kingdoms of Life

As large land animals, we are well familiar with the biodiversity of macroscopic plant and animal life on land, the flowers and trees, mammals and birds, around us. While Darwin in 1837 first speculated on the relationships between different organisms, the German scientist Ernest Haeckel in 1866 recognized that microbes did not fit into the Kingdoms of Plants nor Animals. To accommodate them in his tree of life (Fig. 2) he newly created the Kingdoms of *Protista* (covering protozoa, algae, slime moulds) and *Monera* (from *mono*, one; later called Bacteria or Prokaryota). In 1959 the botanist Robert Whittaker further separated the kingdom Fungi from the Plantae, based on their distinct nutritional modes. Finally, in 1977 Carl Woese [13], based on increased availability of genetic sequences, further separated the *Bacteria* into *Archaea* (the most ancient 'bacteria') and *Eubacteria* (including the cyanobacteria or blue-green algae). Until the introduction of molecular techniques, bacteria had to be grown in culture in the laboratory and be identified using morphological and biochemical characters. Bacteria exhibit very limited morphological complexity, and less than 1% can be cultured. The advent of rapid gene isolation technology afforded by the polymerase chain reaction (PCR) and automated DNA sequencing capabilities opened new windows into microbial diversity. It also soon became clear that one can recover rRNA genes from environmental samples without the need to culture the microbes. Finally in 2005, the introduction of still faster next generation DNA sequencing technology eclipsed our knowledge of microbes by at least 100-fold. *Archaea* for example are now also known from the gastrointestinal tracts of half of the human population (e. g., *Methanobrevibacter smithii*). Viruses tend to have often been excluded from the cell-based tree of life [14,15]. While they contain genetic material in the form of DNA or its sister molecule RNA, they must infect host cells and take over its machinery before they can reproduce copies of themselves to burst into life and synthesize proteins. Of the 23 main divisions of life we recognize today, only plants, animals and some fungi are large enough to be seen with the naked human eye. Our planet is run by microbes.

 | HTTPS://DOI.ORG/10.1515/9783112228265-003

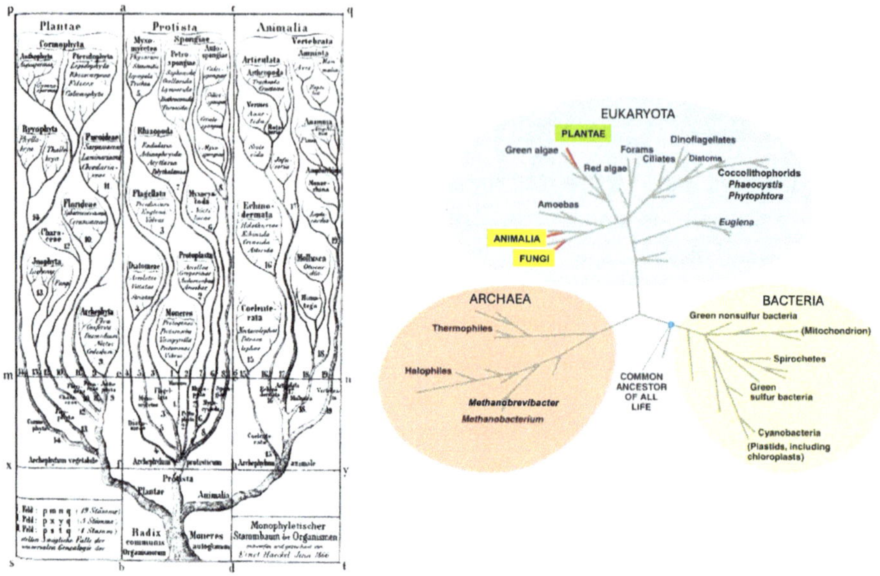


FIGURE 2 Left. Haeckel's (1866) early tree of life positioned the higher plants and higher animals at the crown of evolution but for the first time also recognized the existence of a third Kingdom, the Protista. Right. A radically different view of the phylogenetic tree of life in the 2000s, based on molecular sequences. Shown are the three domains Archaea, Bacteria (including the sulfur bacteria from which originated the mitochondria, and the cyanobacteria from which originated the chloroplasts) and the Eukaryota (with a true nucleus). The latter include the Kingdom of Protists such as diatoms, dinoflagellates and ciliates, Kingdom of Fungi, Kingdom of Plants and Kingdom of Animals. In terms of genetic diversity, life on our planet is dominated by microbes, with plants and animals and humans representing very small branches only.

# PART 1

# Cyanobacteria

CHAPTER 4

# Lungs of the Planet: First Cyanobacteria

## First Cyanobacteria to Produce an Oxygen Atmosphere

Geologists long postulated that the abundant water on our planet was formed as a result of volcanic activity during which hydrogen and oxygen fused together in the Planet's hot interior and rose to the Earth's surface as steam. Freshwater accumulated in ponds, lakes and streams while the dissolution of salts contained in rocks and sediments led to the formation of the oceans. Starting 3.5 billion years ago, violent chemical reactions in our gaseous atmosphere of nitrogen, carbon dioxide and methane, precipitated a rich biochemical soup in our oceans, in which formed amino acids – the building blocks of proteins – and nucleic acids – the building blocks of the genetic material DNA. An alternative theory, advocated by British astronomer Fred Hoyle of "Big Bang" fame, is that life on Earth has an extraterrestrial origin. Recent radioisotope evidence [16] that the age of water in our oceans is older than the age of our solar system adds new credibility to this idea. Under that alternative scenario Earth formed in a region of space that contained water in some form, or the water on Earth was delivered post-formation by other bodies in the solar system.

The first self-reproducing life forms to evolve were bacteria which possess no membrane-bound nucleus and no organelles such as chloroplasts or mitochondria. Most were colorless methane and sulfur-producing bacteria relying on chemosynthesis for growth. These early life forms – now called Archaea – still dominate the ecosystems of the hydrothermal vents in our present-day deep seas. Later forms such as the cyanobacteria, often incorrectly referred to as *blue-green algae*, newly evolved the green pigment chlorophyll to trap sunlight for photosynthesis. In this process carbon dioxide from the atmosphere reacts with water to synthesize sugars which build cell material, while oxygen is released as a by-product. Until the evolution of photosynthesis, life on Earth was driven by anaerobic processes (without oxygen), but the advent of aerobic metabolism allowed eighteen times more energy to be generated, thus paving the way for the emergence of more complex multicellular, eukaryotic organisms [17]. The oldest microfossil interpreted with certainty as a cyanobacterium is *Eoentophysalis belcherensis* originating from 1.8 billion years old deposits from Hudson Bay, Canada [18]. This species is morphologically almost identical to the modern day coccoid terrestrial cyanobacterium *Entophysalis granulosa*. The first appearance of oxygen in our planet's atmosphere – the so-called Great Oxidation Event – is marked by the worldwide occurrence of banded iron formations (Fig. 3) in rocks deposited between two to three billion years ago. This resulted from the oxidation and subsequent precipitation of abundant iron

 | HTTPS://DOI.ORG/10.1515/9783112228265-004

dissolved in ancient seawater. The lucrative iron ore mining industry in Australia, China, Brazil and India all capitalize on this geological event. Much of the oil that we now use to drive our cars began millions of years ago when the sun shone on plankton algae drifting in prehistoric seas and produced, through photosynthesis, minute globules of oil reserves within these microscopic cells. The global warming crisis we humans are faced with today is caused by humans burning every year the equivalent of 1 million years of plankton oil production. Oxygen production by photosynthetic cyanobacteria has also been responsible for the creation around our planet of a protective ozone shield, screening out harmful ultraviolet radiation and thereby allowing for the development of more complex life forms. Every second breath of oxygen we humans inhale originates from past and present photosynthesis by microscopic ocean algae. Photosynthesis is undoubtedly the single most metabolic innovation in the history of life on our planet, and it was invented by cyanobacteria not by land plants. It would be difficult to pinpoint any other microbes that had a more profound lasting impact on our planet.

FIGURE 3 Top: Banded iron formations reflecting the Great Oxidation Event, resulting from the advent 3.5 billion years ago of cyanobacterial photosynthesis. Original photograph from specimen held at Artis Groote Museum in Amsterdam. Bottom: Scanning micrograph of extant filamentous cyanobacteria *Nodularia spumigena* from an Australian estuary.

CHAPTER 5

# *Prochlorococcus:* Ocean's Smallest Photosynthetic Cell

## Most Abundant Photosynthetic Cyanobacterium in the World Oceans Today

The most widespread photosynthetic organism in the world ocean, estimated to contribute 13 to 48% to global carbon fixation, was only discovered in 1986 from the Sargasso Sea simply because it was so minute (0.6 µm diameter) that it required medical flow cytometry technology to be visualized. Initially, its green color was interpreted to signal affinity with primitive green algae such as *Prochloron*. Through the work of Massachusetts Institute of Technology Sallie Chisholm [19] we now know that this microbe is a cyanobacterium *Prochlorococcus* (meaning "primitive green berry') with unique light-harvesting pigments (divinyl derivatives of chlorophyll a and chlorophyll *b*) but lacking phycobilisomes and the typical chlorophyll *a* typical of other cyanobacteria. While the genome of *Prochlorococcus* is quite small, this organism occurs as very numerous strains, some are adapted to high light, others low light. Almost every strain that has been genetically sequenced appears unique, and when taken all together (called a pan genome) represents four times the total of the human genome. As a result, what initially was considered a single species, can be dominant (100 million cells per liter) in tropical and subtropical low nutrient waters such as open ocean gyres, and thrive down to 110 to 150 m depths. In 2020 *Prochlorococcus* was further differentiated into five separate genera including *Eurycolium, Prolificoccus, Thaumococcus* and *Riococcus* [20]. Fast growing microbes evolve at much higher rates than large land plants, and what at first glance may look like a primitive organism commonly hides an astonishing genetic diversity and physiological plasticity.

 | HTTPS://DOI.ORG/10.1515/9783112228265-005

CHAPTER 6

# *Trichodesmium* "sea sawdust"

## Most Widespread Cyanobacterial Bloom in Tropical Oceans

*Trichodesmium erythraeum* is a tropical marine cyanobacterium producing floating colonies made up of straight filaments oriented parallel in bundles. At the start of the bloom the colonies appear throughout the water column, but during late bloom stages the strong gas vacuoles cause a massive rise of the cyanobacteria to the surface layers. These can then appear as yellow–grey (early bloom) or reddish brown (late bloom) colored windrows spreading over up to 40,000 $km^2$. The Yellow Sea and Red Sea are named for such blooms. *Trichodesmium* red tides ('sea sawdust') were observed as early as 1770 during Captain Cook's voyage through the Coral Sea who mistook them for coral spawn [21]. This species can be a nuisance to swimmers on Australian beaches, but harmful effects on fish are seldom observed except in sheltered bays where the decaying bloom may generate anoxic conditions (e.g., in India [22]). An unusual death of corals caused by the decomposition of masses of *Trichodesmium* driven ashore by the wind has been reported in New Caledonia. Furthermore, there are reports of breathing difficulties from people in Brazil who have been near blooms of *Trichodesmium* ('Tamandare fever' [23]). There is no evidence of a relationship with pollution. In fact, differentiated cells within the center of the colony are capable of fixing atmospheric nitrogen, which allows the species to thrive under nutrient-impoverished oceanic conditions, where they readily outcompete other microalgae. Wave action can break up the bundles and inactivate the nitrogen-fixing nitrogenase enzyme, which is why calm seas are a prerequisite for blooms. *Trichodesmium* plays a key role in fertilizing nutrient poor tropical waters with atmospheric nitrogen, accounting for almost half of the nitrogen fixation in marine systems globally. Related nitrogen-fixing cyanobacteria such as *Chlorogloea*, *Nostoc*, *Pseudoanabaena* are critical for natural fertilization of rice paddies [24]. This is not unlike the role that *Rhizobium* bacteria play in fertilizing agricultural fields planted with legumes (see soil bacteria). By comparison, for humans to produce ammonia fertilizer from atmospheric nitrogen (using the Haber-Bosch chemical process) we need to heat source materials to 500°C under 300 times atmospheric pressure, while also generating significant unwanted greenhouse emissions.

 | HTTPS://DOI.ORG/10.1515/9783112228265-006

CHAPTER 7

# Cyanobacterial Blooms and Drinking Water

## Toxigenic Cyanobacteria Increasingly Threaten Freshwater Drinking Water Supplies in a Warming World

Harmful cyanobacterial blooms in freshwater pose human health threats through contamination of drinking and recreational waters (Fig. 4). The earliest written record of such toxic bloom dates back to 1878 when cattle deaths were associated with the brackish water cyanobacterium *Nodularia spumigena* in Lake Alexandrina in South Australia [25]. Francis 1878 in a note in the journal Nature reported that

> a 'conferva' indigenous and confined to the lakes had been produced in excessive quantities, so much as to render the water unwholesome. It is swallowed by cattle when drinking. This acts poisonously and rapidly causes death. Sheep from 1 to 6 or 8 hours, horses 8 to 24 hours, dogs 4 to 5 hours, pigs 3 or 4 hours.

In 1979, 140 children and ten adults of the aboriginal population of Palm Island off Queensland, Australia, received hospital treatment after a *Raphidiopsis* (*Cylindrospermopsis*) *raciborskii* bloom in their drinking water reservoir [26]. A widespread problem occurred in central Australia in November–December 1991 when a bloom of *Dolichospermum* (*Anabaena*) *circinale* covered 1,000 km of the Darling-Barwon river system and killed an estimated 10,000 livestock and required army emergency water supplies to be brought in for several townships [27]. More insidiously, 55 human fatalities occurred in Brazil in 1996 when microcystins from *Microcystis panniformis* contaminated water was inadvertently offered to patients in a hemodialysis clinic [28]. Chronic toxicity from microcystins in drinking water has also been associated with higher than normal incidences of liver and colorectal cancer in Chinese townships unknowingly serving contaminated drinking water. On August 2nd, 2014, residents of Toledo, Ohio, USA, awoke to urgent warnings not to drink or use their tap water. Half a million people were unable to drink their water, cook with it, or brush their teeth. The cause: a massive *Microcystis aeruginosa* bloom in eutrophic Lake Erie. Toxic blooms have also been linked to deaths of dogs licking their coats after swimming in New Zealand ponds with *Phormidium autumnale*, and are also of concern when transmitted via leafy vegetables irrigated with contaminated water. In most Western countries drinking water supplies are now routinely monitored using sensitive chemical analytical methods for cyanotoxins such as cylindrospermopsins, microcystins, nodularin and saxitoxins. Nutrient pollution by phosphorus-containing household cleaning products and agricultural

 | HTTPS://DOI.ORG/10.1515/9783112228265-007

fertilizers together with global warming are responsible for warmer summers and increasing cyanobacterial bloom problems [29].

*Not to be confused*: Drinking water can also be contaminated with protozoa such as *Giardia* and *Cryptosporidium*. Giardiasis is a bowel infection caused by a tiny protozoan parasite called *Giardia lamblia*, also known as *Giardia intestinalis*. This is a single-celled protist that can attach itself in large numbers to the wall of the human bowel and interferes with the body's natural absorption of nutrients. In August 1998, following heavy rainfall events, the drinking water supply of metropolitan Sydney, Australia, was found to be contaminated with both *Giardia* and *Cryptosporidium* cysts. Health warnings were issued leading to comprehensive boil water alerts.

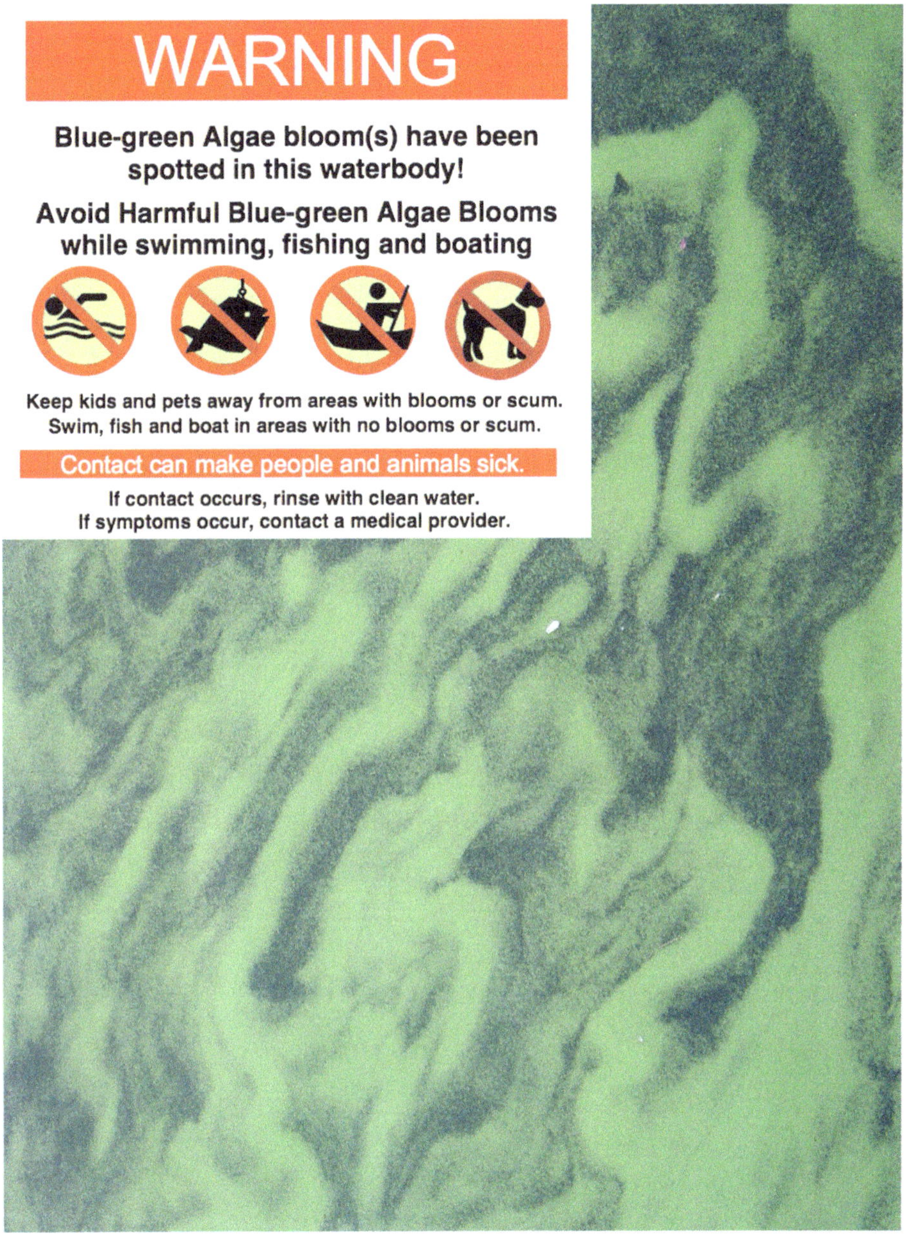


FIGURE 4 Cyanobacterial bloom of *Microcystis aeruginosa* from a Dutch canal. Inset: Toxic bloom warning poster from New York State Department of Health.

CHAPTER 8

# *Spirulina* and the Aztecs

## One of the First Applications of Algal Biotechnology

During the 1400s and 1500s the Aztecs human population of Latin America and the Kanembu people of Chad in Africa were known to harvest *Spirulina* cyanobacteria (now called *Arthrospira*) from coastal lagoons, collect them in clay pots, drain out the water through bags of cloth, and spread out the algae in the sun for drying. When dry, women took these algal cakes for sale at the local market. The product Dihé was eaten in 70% of their meals. Pregnant women believed that its dark green color screened their unborn babies from the eyes of sorcerers while *Spirulina* was also applied externally for treating certain diseases. It was not until the 1960s that *Spirulina* was rediscovered by western science as a "future food source" due to its exceptionally high protein content of 70% of its dry weight [30]. Just 20 g of *Spirulina* will provide an adult human with their daily requirement of vitamin $B_{12}$ as well as 70% of their requirement for vitamin $B_1$ and 50% of vitamin $B_2$. Today, *Spirulina* is being produced in more than twenty-two countries for use in nutritional supplements, high end cosmetics, water purification and algal biofuels. In 2022 the global *Spirulina* market was valued at USD 542 million. NASA (North Atlantic Space Agency) is even exploring *Spirulina* as a food source for astronauts.

 | HTTPS://DOI.ORG/10.1515/9783112228265-008

# PART 2

# Algae

CHAPTER 9

# Microscopic Algae: Grasses of the Sea

Microscopic plankton algae range in size from 1 µm to 2 mm. They live a floating existence in the upper 200 m of the ocean. Like land plants, they require sunlight, carbon dioxide, water and dissolved nutrients phosphate, nitrate, silicate for photosynthesis and cell growth. These single-celled ocean plants are collectively known as *phytoplankton* (from the Greek *phyto* meaning 'plant' and *planktos* 'made to wander'). Algae have no need for the complex support structures found in land plants. They do not have readily distinguishable flower, leaf, stem and root systems, and therefore are classified as a primitive form of plant life or *algae* (singular *alga*). These early aquatic life forms, which first evolved 3.5 billion years ago, represent a much greater genetic diversity than the land plants and animals which are relative late comers and appeared only 400 million years ago. Charles Darwin already recognized in 1859 in his book "*The Origin of Species*" that terrestrial forests do not maintain nearly as many species as those "which swarm in the bosom of the sea". This motivated the maverick biologist Craig Venter, who in 2000 raced the US Government to complete the Human Genome Project, to genetically map the living world in its totality by sequencing the largely untapped reservoir of marine microorganisms. The French *Tara Oceans* expedition from September 2009 to December 2013 used both sophisticated microscopic (flow cytometry, Flow Cam) and molecular technologies to further pursue this quest [31].

The color of the ocean is forever changing and varies according to the location and the seasons [32]. Open ocean waters are usually blue, but coastal waters tend to be green or brown. These differences are due to variations in the abundance of microscopic plankton algae (Fig. 5). One single liter of seawater can contain as many as one million algal cells belonging to up to one hundred different species. Furthermore, whereas plant life on land is dominated by a single evolutionary branch of organism, the higher plants, plant life in the oceans includes representatives from as many as fourteen higher taxonomic units called divisions or phyla [33]. These organisms show an immense diversity of form as well as cellular structure and pigmentation ranging from red, blue, green, brown or yellow, all of which are adaptations to the ocean environment. Land plants are always grass-green in color, but golden-brown pigments predominate in the oceans. Just as a medical practitioner takes blood samples to monitor a patient's health, plankton biologists monitor the environmental health of water ways by looking at algal species.

 | HTTPS://DOI.ORG/10.1515/9783112228265-009

Bob Andersen from the US Provasoli-Guillard Center for Culture of Marine Phytoplankton calculated that if all the algal cells in the oceans could be packed into a plank, 7 cm thick, 30 cm wide, it would stretch 386,000 km from the Earth to the Moon [34]. While the biomass of microscopic algae may appear insignificant compared to that of the forests on the land, a significant proportion of woody land plants are inedible. In contrast, the oceans produce a new plank of similar dimensions each day which is consumed in its entirety by food chains leading to fish which are edible to humans. While microscopic ocean algae reproduce themselves roughly once a week, the world's land plants on average need ten years to reproduce [35–37]. If we translate the oceans into the language of economics, its resources and services to humanity are worth a staggering USD 21 trillion per year, which exceeds the value of the total global gross national product.

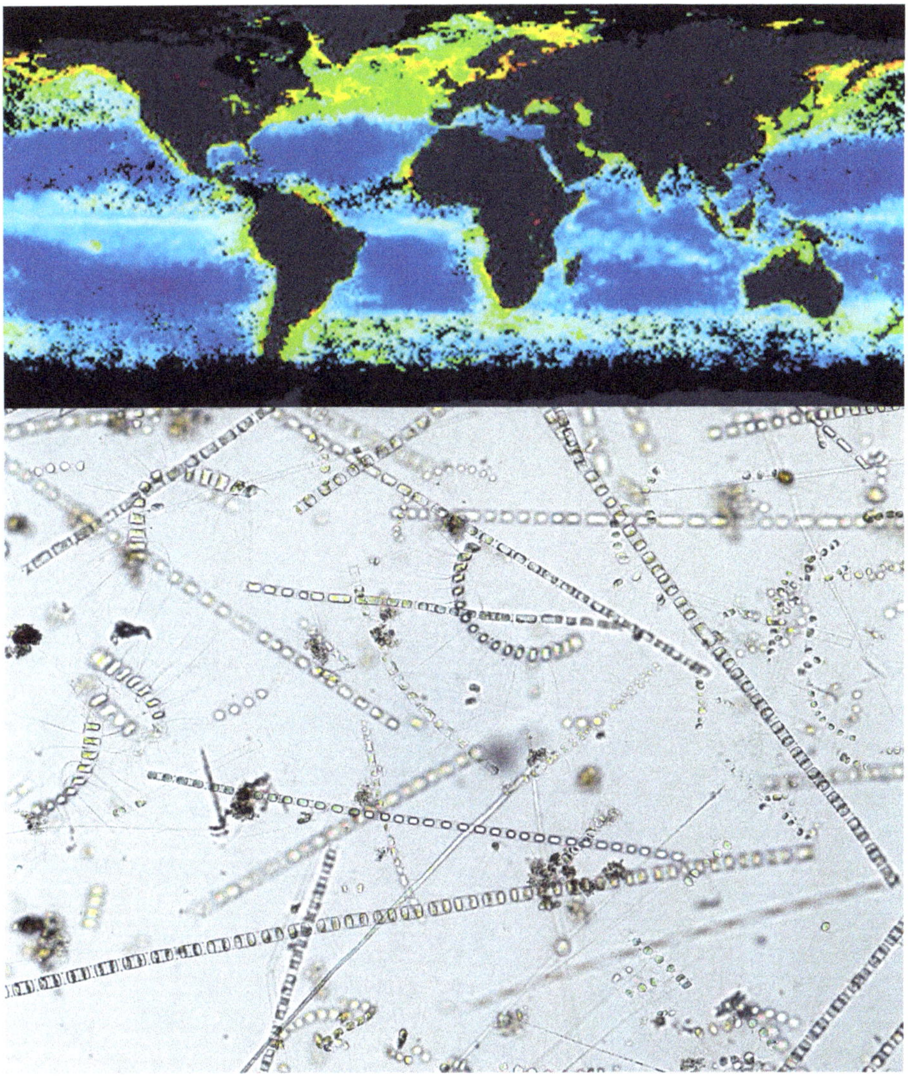

FIGURE 5 Top. NASA satellite image of surface chlorophyll-*a* phytoplankton pigments during the Northern Hemisphere spring diatom blooms (highest concentrations in yellow and orange). Below: Light micrograph of diatom spring bloom in South Australian waters dominated by chain-forming colonies of *Skeletonema* and *Chaetoceros.*

CHAPTER 10

# Coccolithophorid *Emiliania huxleyi* and Gaia

## Inspiration for Gaia Hypothesis. The Model Microalga in Today's Oceans in Terms of Carbon Sequestration

Coccolithophorids (from the Greek *coccus* 'berry'; *lithos* 'stone'; *phorid* 'carrying') are a group of very small plankton algae, referred to as nanoplankton (*nano* means dwarf), mostly 2 to 20 μm in size, which are covered by delicate calcareous scales known as coccoliths [38]. These structures were first discovered in marine sediments in 1854 by the German scientist Gottfried Ehrenberg who thought that they were crystals of inorganic origin [39]. It was not until the turn of the 19th century that living coccolithophorids were discovered swimming freely in the surface layers of the ocean, and their biological origin was recognized. Nobody really knows what value the delicate tapestry of coccoliths holds for the algal cells. Candidate theories include a skeleton function, to protect against grazing animals or virus attack, ballast to regulate buoyancy, a mechanism to prevent calcium poisoning, or a light-scattering function to stimulate photosynthesis. In recent years, dense concentrations of coccolithophorids (up to a million cells per liter) in the North Sea or Jervis Bay, Australia, have occasionally resulted in milky-white seas which are visible by satellite. The ubiquitous spherical coccolithophorid *Emiliania huxleyi* (named after the naturalist Thomas Huxley) is covered with calcareous scales (coccoliths) which are composed of T-shaped elements, radially arranged as in a spoked wheel. These organisms were so abundant in prehistoric times (140 to 65 million years ago, named after them the Cretaceous) that they have created calcareous deposits such as the white chalk cliffs of Dover (Fig. 6). *Emiliania* blooms absorb the greenhouse gas carbon dioxide in the process of photosynthesis, but at the same time produce cloud seeding droplets of dimethyl sulfoxide that reduce sunlight and thereby photosynthesis. This inspired the late James Lovelock's Gaia hypothesis [40]. The story of Gaia began with a question posed by NASA scientists while Lovelock was a consultant at the Jet Propulsion Laboratory in Pasadena, California. That is, how could you tell if a planet such as Mars harbored life? Earth's atmosphere contains oxygen- and methane-reactive gases, that are constantly renewed. That disequilibrium radiates an infrared signal, which Lovelock later described as an "unceasing song of life" in striking contrast to the static Martian atmosphere, composed almost entirely of non-reactive carbon dioxide. Lovelock's Gaia theory states that, for much of the past 3.8 billion years, a holistic feedback system has played out in the biosphere, with life

 | HTTPS://DOI.ORG/10.1515/9783112228265-010

forms regulating temperature and proportions of gases in the atmosphere to life's advantage. Historic periods with dense algal blooms, such as those that created the white cliffs of Dover, have been suggested to have triggered ice ages [41]. This inspired ideas of geoengineering the oceans by iron fertilization to combat greenhouse warming [42]. A moratorium agreement, adopted on 30 May 2008 at the United Nations Convention on Biological Diversity in Bonn, Germany, enforced a complete ban on such projects until scientists better understand the potential risks of manipulating the oceanic food chain.

FIGURE 6 Top: White cliffs of Dover comprised of the calcareous remains of microscopic planktonic coccolithophorids. Bottom: Scanning electron micrograph of the cultured coccolithophorid *Emiliania huxleyi*.

CHAPTER 11

# *Phaeocystis* and Sea Foam

## Sea Foam can Cause Problems for Tourism, Surfers, Fisheries and Aquaculture, and Even Nuclear Plant Cooling Water

In May 2020, five Dutch surfers who knew the sea "like the back of their hand" died after a huge layer of foam in the water disoriented them and hampered rescue efforts. The group ran into difficulties at the northern harbor head of the Scheveningen district of The Hague in the Netherlands. The sea foam that led to their deaths originated from the golden-brown alga *Phaeocystis globosa* [43].

Colonies of these small coccoid and flagellate algae are held together by a mucous-like matrix that protects the cells from grazing and helps them float (Fig. 7). The algae need sunlight and nutrients such as nitrogen and phosphate [44], and a sunny start to May helped the cells grow rapidly. The sea foam off Scheveningen, which the Dutch call "Mei Sop" [May Froth], was four times more abundant than the long-term average in the past ten years. Cloudy weather the day before the tragedy triggered the disintegration of the algae colonies and viral infections caused the cells to burst open. Finally, a force-7 wind and strong seas whipped them up into a large amount of foam against a harbor head before the currents changed and the foam moved toward the beach.

Dense blooms of *Phaeocystis* also occur in the Southern Ocean. When expeditioners shower on board research vessels that employ desalinated seawater, they come out smelling like cabbage due to dimethyl sulfoxide chemicals from *Phaeocystis* (see also coccolithophorids and Gaia). The characteristic "fishy" smell of the sea is attributed to these same sulfur compounds, showcasing plankton's influence on our sensory experience of the ocean. *Phaeocystis* blooms in the North Sea have also been known to alter the migratory path of herring [45], and been associated with mass mussel mortality in the Netherlands in 2001. In late 2014 a dense *Phaeocystis* bloom in Beibu Gulf in China that lasted for two months caused blockage in the cold-water intake system of a nuclear power plant [46]. In classical mythology, the ancient goddess of beauty and love Aphrodite, identified as Venus by the Romans (as in Botticelli's famous 1485 painting), was born from the white foam produced by the severed testicles of Uranus (Heaven) after his son Cronus threw them in the sea.

 | HTTPS://DOI.ORG/10.1515/9783112228265-011

FIGURE 7 Top: Sea foam on the beaches in the Netherlands caused by *Phaeocystis* golden-brown flagellate blooms. Bottom: Light micrograph of cultured cells.

CHAPTER 12

# *Phytophthora infestans* and Irish Potato Famine

## Key Agricultural Pathogen Responsible for Major Human Migration Event from Europe to America

*Phytophthora* resembles a filamentous fungus but today is classified as a slime mould (oomycete) more closely related to brown seaweeds. It caused the Irish potato famine from 1845 to 1852 when one out of eight million Irish people died, and 1.5 million emigrated to the United States. The potato has its origins in South America and was introduced into Ireland in 1589. Because the potato grew easily, even in poor soil conditions, it soon became the food staple of Irish life. When in 1845 the summer weather in Ireland suddenly changed to a very wet and cool climate, *Phytophthora infestans* decimated the potato crops that had limited genetic variability and no resistance [47]. This parasitic microbe continues to devastate potatoes and tomatoes worldwide, costing billions of dollars annually in losses and control costs such as fungicides, genetically engineered crops and certified seed potatoes. The related *Phytophthora cinnamomi* ranks among the 100 world's worst invasive species causing dieback or root rot in Australian and North American forest ecosystems but also has affected food crops such as avocado and pineapple. Other species of this group such as *Phytium* are a nuisance to gardeners as it causes root and stem rot diseases.

*Not to be confused*: The wheat rust *Puccinia graminis* is a true fungus responsible for significant economic losses to cereal crops. Rust was already known in Aristotle's time and is treated with fungicides and the breeding of resistant wheat strains. Other plant pathogens include bacteria such as *Agrobacterium tumefaciens* that causes crown galls. The best way to control this latter disease is to sterilize pruning tools to avoid infecting new plants or soak germinated seeds, seedlings and rootstock in a suspension of the non-pathogenic bacterial competitor *Agrobacterium radiobacter*.

 | HTTPS://DOI.ORG/10.1515/9783112228265-012

CHAPTER 13

# Diatoms: Miniature Architects

> Adhering to roots of the pond-weed (and sometimes separate in the water) I saw many pretty branches, composed of rectangular oblongs and exact squares. (Anonymous, Letter to the Royal Society of London, 1703)

The wondrously shaped diatoms are the best known and most abundant of all the single-celled plankton algae. These jewel-like stars, diamonds, squares and triangles have fascinated amateur microscopists ever since the first genus, *Bacillaria*, was formally described in 1791. Their silica shells or frustules are composed of two parts (*diatoma* means 'cut in half') fitting into each other like a pill-box. It could be said that diatoms spend all their lives in glass houses. These organisms range in size from 2 to 4,000 micron in diameter. Diatom shells are found in a variety of shapes and may be circular, boat-shaped, square, oblong, triangular, elliptical or polygonal (Fig. 8). The intricate ornamentation of the silica wall, upon which diatom taxonomy is based, has resulted in more than 100,000 species being described [48]. These tiny plants can occur as single cells or more often in long chains. Individual cells are connected into rod-shaped or necklace-like chains via interlocking siliceous processes, or mucus (chitan) strands exuded from tube-like strutted or labiate processes. Chain formation by these species assists in flotation by increasing form resistance, and acts as a defense mechanism against predation by plankton animals.

On the basis of valve symmetry, two separate orders are distinguished, the pill-box type or centric diatoms with concentric or radial symmetry around a point or points, and the cigar shape or pennate diatoms with symmetry around a line. The silica layer has a pattern of depressions or cavities. Regularly arranged holes that appear as small dots in the light microscope are called areolae, while rows of areolae and long narrow cavities are called striae. In the scanning electron microscope, it appears that areolae are pit-like depressions partly closed from the outside or inside with a thin perforated membrane (cribrum). Hexagonal honeycomb-like areolae are especially prevalent in diatoms. The various systems of hollows allow for maximum mechanical strength without making the walls too heavy and without requiring much silica for construction. The places where you are likely to find the golden-brown pigmented diatoms are floating in the plankton of the oceans (dominated by centric diatoms), but also in the plankton of lakes and estuaries, especially where nutrients are brought to the surface by wind or currents and light is available for photosynthesis. Plankton diatoms are the main food for small plankton animals (zooplankton), such as copepods and shrimp-like krill, which use their hairy legs to catch the algae and in turn are eaten by small fish. Nutrient-rich areas, such as the

 | HTTPS://DOI.ORG/10.1515/9783112228265-013

Peru upwelling system or Southern Ocean, support rich diatom plankton and form the basis of the richest fisheries in the world's oceans. Other diatoms live on the seabed of shallow estuaries, accounting for the brown film on tidal mudflats (mainly pennate diatoms). Just after the tide goes out, these mudflat diatoms glide to the surface for a sun bath. They know precisely when to burrow back into the mud before the return of the tide. Diatoms also form a brown coating under polar ice floes or catch a ride on whale skins or bird feathers. On land, they live under moist conditions in top soil, attached to moss, tree trunks or even brick walls.

> What happens to a diatom in the upper, sunlit strata of the sea may well determine what happens to a cod lying on a ledge of some rocky canyon a hundred fathoms below. (Rachel Carson 1951 [49])

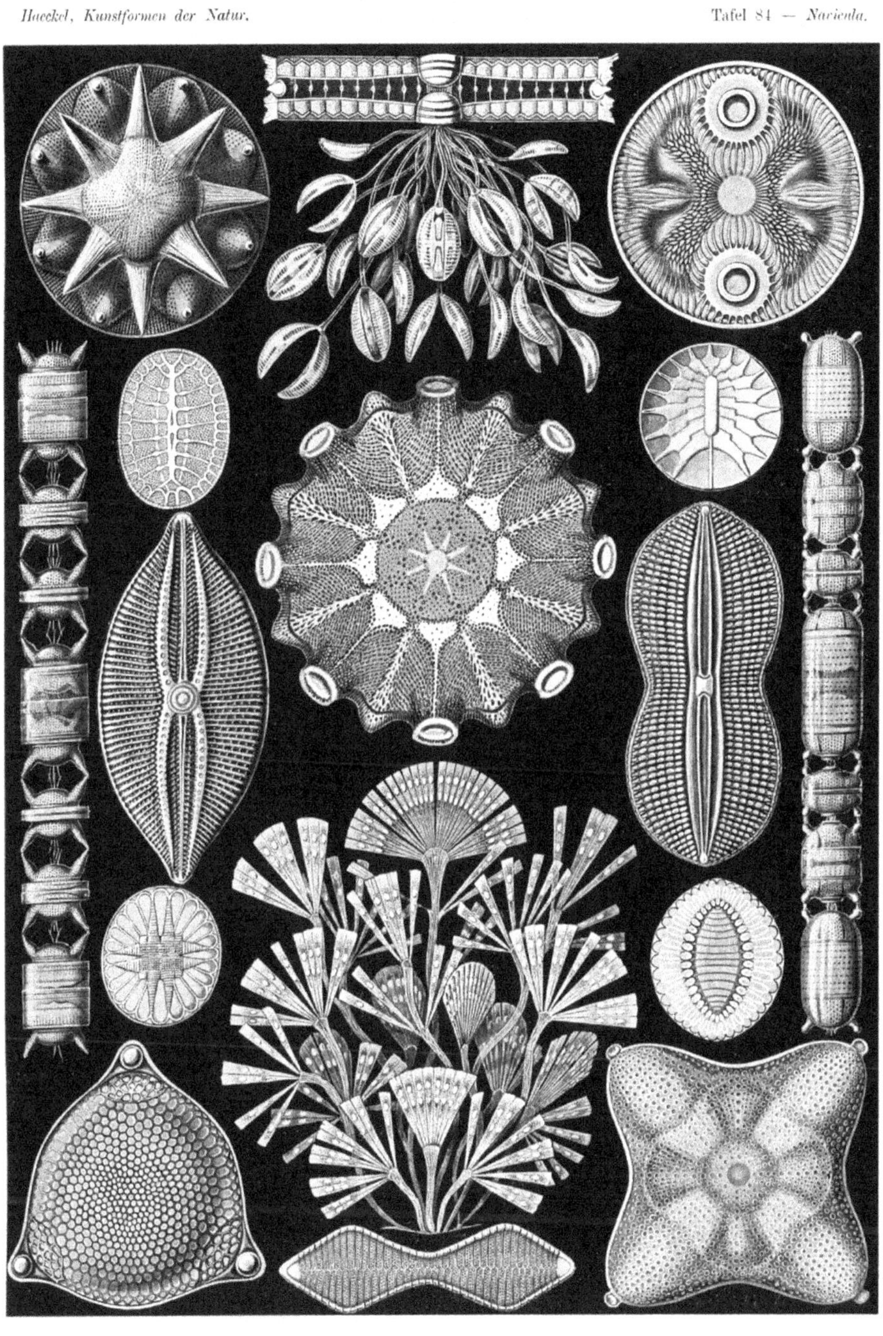


FIGURE 8 Diatoms. Plate 84 from Haeckel, E. 1899–1904. *Künstformen der Natur.*

CHAPTER 14

# Blue Diatom *Haslea ostrearia:* Oysters Fit for Royalty

## Blue Diatoms Create a Unique Oyster Delicacy

For centuries French food connoisseurs have celebrated "*fine de claires vertes*" oysters from the Marennes-Oléron region on the French Atlantic coast. These oysters are conditioned in knee-deep rectangular salt ponds, called 'claires', for a minimum of one month, during which they fatten and take on a sweeter, fruitier flavor from the unique microalgae inhabiting these ponds. One such species is the diatom *Haslea ostrearia* that produces an unusual water soluble blue pigment, called marennine after the region where it was first found. This pigment stains the oyster gills an extraordinary shade of turquoise blue-green. Such oysters were favored by French royalty as early as the 1700s, and in France oyster growers attract twice the price for blue oysters than for their regular stock. While this phenomenon has occasionally been observed in Australia, there local oyster growers have been mostly unaware of what caused the coloration. French scientists at the University of Le Mans in France led by Jean-Luc Mouget [50] have significantly advanced our understanding of this phenomenon. The blue pigment is harmless to humans, but has antioxidant, antibacterial and antiviral properties, which may give blue oysters an advantage to cope with environmental stressors such as heat waves and viral infections. This humble diatom offers oyster growers a market and branding opportunity to work with restaurants and distributors for the benefits of culinary delights for shellfish consumers.

 | HTTPS://DOI.ORG/10.1515/9783112228265-014

CHAPTER 15

# *Didymosphaenia geminata* or "rock snot"

## Nuisance Diatom Fouling Problem in Freshwater Streams

While many diatoms float in the water, others attach as a brown coat to surfaces such as buoys, rocks and ships. The large bottle-shaped diatom *Didymosphaenia geminata* (*Didymo* for short) produces thick, branching stalks to attach itself to rocks in freshwater streams, earning it the name "rock snot" (Fig. 9). Historically its distribution was in cold waters of North America and Europe, until its sudden appearance in October 2004 in the South Island of New Zealand, putatively introduced as stowaways on the moist waders of trout fishermen [51]. Masses of cells and stalks can extend over 1 km and persist for months creating the appearance of fiberglass insulation, tissue paper, "rock snot", brown shag carpet or sheep skin covering the stream bed. Strict quarantine regulations of cleaning fishing gear with chlorine were newly introduced in New Zealand to halt further spreading. Air travellers entering Australia from New Zealand now have to complete customs declarations that they have not visited freshwater streams and not carry *Didymo* stowaways on their fishing gear or waders. A similar problem in Japan was recognised in 2006 caused by the stalk-forming diatom *Cymbella janischii* of North American origin [52].

 | HTTPS://DOI.ORG/10.1515/9783112228265-015

FIGURE 9 Top: Brown "rock snot" from the Buller River, New Zealand, caused by the growths of the diatom *Didymosphaenia geminata*. Bottom: Scanning electron micrograph of the large bottle-shaped diatom cells.

CHAPTER 16

# *Pseudo-nitzschia* and Hitchcock's *The Birds*

## Cause of Seafood Quality Problems Affecting Both Humans and Marine Mammals

On 18 August 1961 the Californian newspaper *Santa Cruz Sentinel* reported how thousands of crazed seabirds pelted homes, unable to fly and regurgitating anchovies. This incident inspired Alfred Hitchcock to produce his famous 1963 thriller *The Birds,* albeit taking the artistic liberty to use black crows instead of seabirds (Fig. 10). A repeat event occurred on the Californian coast in 1991, this time affecting brown pelicans and sea lions that had been feeding on anchovy which in turn had consumed large quantities of the diatom *Pseudo-nitzschia australis.* Some 30 species of this diatom genus can produce chemical substances (domoic acid) which are capable of poisoning human consumers of shellfish, as occurred in an outbreak of Amnesic Shellfish Poisoning in Prince Edward Island, Canada, in 1987 that hospitalized 105 people and caused three fatalities. That event was caused by the related diatom *Pseudo-nitzschia multiseries* and allowed Canadian scientists [53] to identify the chemical agent domoic acid. Victims suffered permanent brain damage leading to short-term memory loss, hence the designation as amnesic shellfish toxins. While no such human poisoning have recurred so far, adverse impacts on wildlife such as seabirds, sea lions, manatees, dolphins and whales may be more common. Climate change is driving blooms into new areas such as Alaska and the Bering Sea where native fishermen and marine mammals have not been previously exposed to this problem for thousands of years. Once an obscure microorganism, only identifiable by a handful of scientists using electron microscopes and molecular sequencing, this species now is the subject of global plankton and toxin monitoring programs to protect human and ocean health.

 | HTTPS://DOI.ORG/10.1515/9783112228265-016

FIGURE 10 Hitchcock's movie "*The Birds*" was based on an incident in California in 1961 when crazed seabirds pelted homes, unable to fly and regurgitating anchovies, which had been contaminated with toxins from a *Pseudo-nitzschia* diatom bloom.

CHAPTER 17

# Diatom Ooze and Alfred Nobel's Dynamite

## Funding for the Nobel Prizes

Sediments containing large numbers of well-preserved diatom cell walls are formed by the sinking of living diatoms from the surface water. With burial and compaction, the diatom ooze is transformed into the siliceous sedimentary rock termed diatomite (Fig. 11). Diatomite is easily crumbled into a fine white to light yellowish gray powder ranging from 10 to 200 μm in particle size. The powder is abrasive and has a low density as a result of its high porosity. The typical chemical composition of diatomite is 80 to 90% silica, with 2 to 4% alumina from clay minerals, and 0.5 to 2% iron oxide [54]. The Greeks and Romans used it as building material, for example in the dome of St Sophia in Constantinople. Alfred Nobel in 1876 patented the properties of diatomite as a carrier for nitroglycerin, which then could be transported and handled more safely than on its own. The fortune he made during the late nineteen century now funds the Nobel prizes awarded by the Swedish Academy of Sciences. Diatomite is no longer used in explosives today. The major use of diatomite is instead in filtration including the purification of beer, liquors, and wine, the cleansing of greases and oils, production of cement, removal of microbial contaminations, mechanical insecticide, absorbent, matting agent, a fine abrasive such as in tooth paste or shoe polish, and a filler in paints. Diatom oozes composed of different species with differing porosities serve different industrial purposes. Identification of diatom frustules has also been used to trace the origin of clay used for Neolithic pottery [55], to track the migratory path of seabirds, turtles and whales, and as forensic evidence of human fatalities from drowning. Diatom frustules have even been detected adhering to the outside of the US space station.

 | HTTPS://DOI.ORG/10.1515/9783112228265-017

FIGURE 11 Scanning electron micrograph of diatom rich sediments ("diatomaceous ooze") from Adelie Shelf, Antarctica.

CHAPTER 18

# Dinoflagellates: Masters of Chemical Warfare

## Producers of a Wide Range of Chemical Warfare Compounds of Human Health Significance

Dinoflagellates have two whip-like flagella (*flagellus* means 'to whip') that help them to swim up or down in the water column and to choose optimal conditions of light, temperature, salinity and nutrients. A ribbon-shaped transverse flagellum encircles the cell in a horizontal girdle groove and a smooth longitudinal flagellum projects from a vertical sulcus groove. The combined action of the two flagella (singular is flagellum) creates a spiraling course of movement (*dino* means 'whirling'). The cells are often encased by an armor of cellulose plates that fit together like jigsaw pieces called plate tabulation. The number, shape and arrangement of these plates are important for species identification (Fig. 12). Dinoflagellate species exhibit a marked differentiation of the cellulose wall which can be ornamented with spines, horns, fins, ribs, and reticulations. This increases the cell's surface resistance and thereby facilitates flotation. In addition to the armored dinoflagellates, there are also hundreds of so-called naked ('gymnodinoid') species which are bound only by a membranous covering.

About half of the 2,000 living species of dinoflagellates are photosynthetic and behave as plants, but many others feed on algae and bacteria, and behave like animals, or even better most can do both (called mixotrophy). Dinoflagellates are therefore being studied by both botanists and zoologists, in addition to micropaleontologists who study their fossilized resting cysts. Due to an initial lack of communication between these three disciplines, some species are known by as many as three different names, a problem which is gradually being resolved. Dinoflagellates such as *Pyrodinium bahamense* in Puerto Rico and *Noctiluca* in Australia, have become tourist attractions through their ability to produce striking night spectacles of bioluminescence. Other species receive negative publicity through their ability to produce potent toxins that can kill fish (for example *Karenia brevis* in Florida) or cause illnesses in humans such as Paralytic Shellfish Poisoning by *Alexandrium catenella* and *Gymnodinium catenatum*, or Ciguatera Poisoning by *Gambierdiscus toxicus*. Dense surface blooms of both toxic and nontoxic dinoflagellate species can produce spectacular red or brown water discolorations, known as *red tides* (Fig. 13). These phenomena are capable of killing fish and marine invertebrates through the generation of anoxic conditions as already described Exodus 7:20–1

> All the waters that were in the river were turned to blood. And the fish that were in the river died; and the river became foul, and the Egyptians could not drink of the water from the river.

 | HTTPS://DOI.ORG/10.1515/9783112228265-018

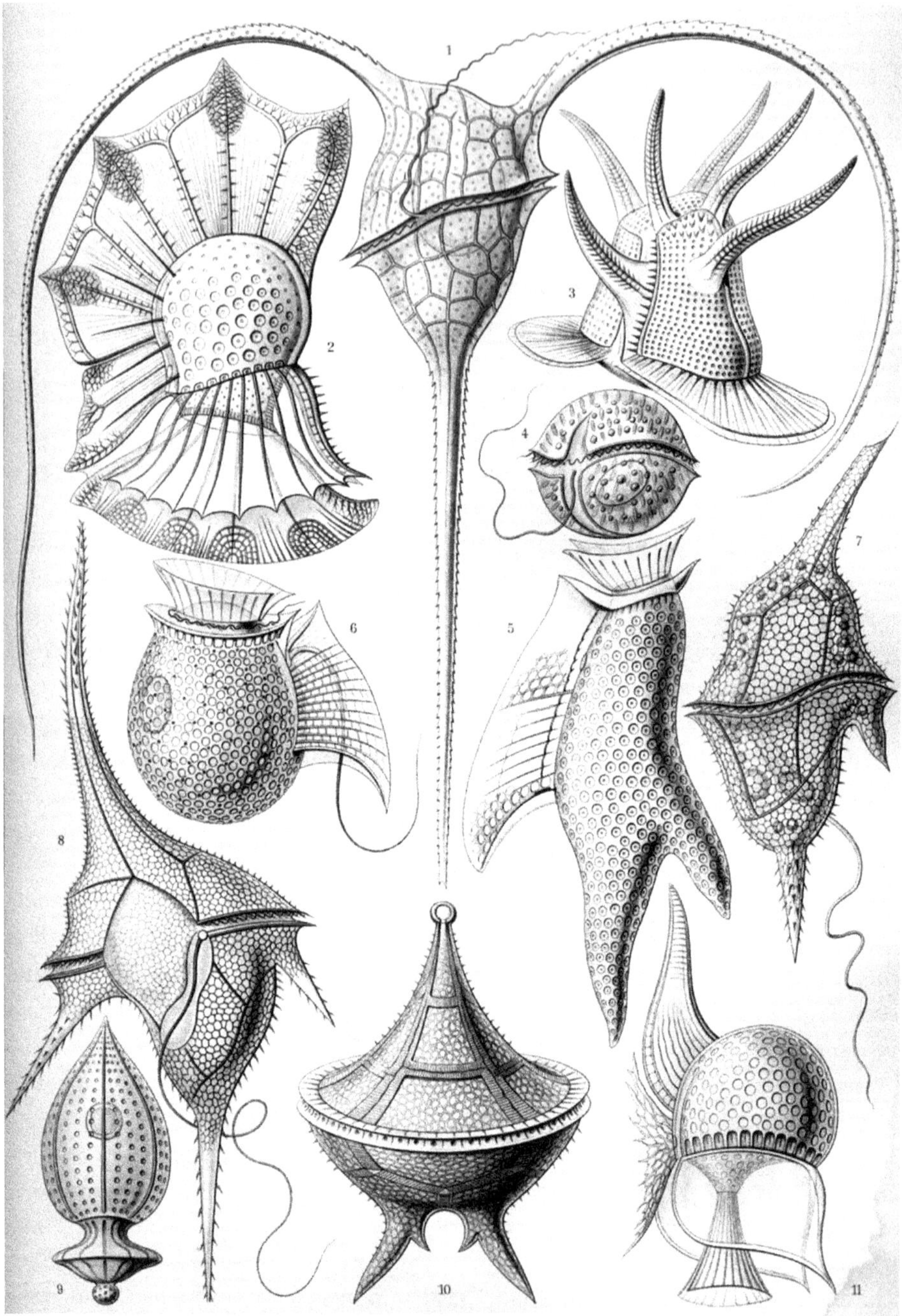


FIGURE 12 Dinoflagellates. Plate 14 from Haeckel, E. 1899–1904. *Künstformen der Natur*.

CHAPTER 19

# *Alexandrium* and Captain Vancouver's Poison Cove

## Major Cause of Human Fatalities from Shellfish Consumption

While most microscopic algae provide excellent food for filter-feeding shellfish and larvae of crustaceans and finfish, other so-called Harmful Algal Blooms (HABs) can have negative effects, causing severe economic losses to aquaculture, fisheries and tourism. Of greatest concern to human society are blooms of toxic species that cause illness and death of fish, seabirds and mammals via toxins transferred through the food web. One of the first recorded fatal cases of human Paralytic Shellfish Poisoning (PSP) after eating shellfish contaminated with dinoflagellate toxins occurred in 1793, when Captain George Vancouver and his crew landed in British Columbia in an area now known as Poison Cove [56]. He noted that it was taboo for local Indian tribes to eat shellfish when the sea became bioluminescent due to plankton blooms. The causative dinoflagellate *Alexandrium catenella* was first formally described in 1936 and the neurotoxin it produces (saxitoxin named after the butter clam *Saxidomus*) chemically characterized in 1975. Doses of 1 mg cause moderate symptoms in humans (tingling sensations around finger tips and lips) but doses of 10 mg can be lethal, resulting in death from respiratory paralysis. When in 1960 a United States U-2 spy plane was shot down by the Soviet Air Defence Forces while conducting photographic aerial reconnaissance deep inside Soviet territory, American pilot Gary Powers was carrying suicide capsules containing saxitoxin [57]. To prevent human poisonings from shellfish consumption, the United States Food and Drug Administration (US FDA) introduced a compulsory monitoring program in 1937 whereby seafood products are regularly tested in accredited laboratories using mouse bioassays. Shellfish containing more than 0.8 mg toxin/kg are deemed unsuitable for human consumption and prohibited from sale. Due to increasing concern over animal ethics, algal toxins are now analyzed using sophisticated chromatographic analytical methods. Similar to rapid antigen tests for COVID, rapid test kits can provide an on-site or on-board qualitative yes/no result within 20 minutes, allowing producers to make on-farm harvest decisions prior to seafood product processing and transport, thus reducing their business risk [58]. Of the global total of 3,800 human Paralytic Shellfish Poisonings during the 1985–2018 period, the largest number occurred in the Philippines, a country strongly dependent on aquaculture for human food protein, with 2,555 poisonings recorded between 1983 and 2013 of which 165 were fatalities, predominantly caused by the highly toxic tropical dinoflagellate *Pyrodinium bahamense*.

 | HTTPS://DOI.ORG/10.1515/9783112228265-019

FIGURE 13 Toxic shellfish warning poster by the Tasmanian Government Department of Health and Human Services, against the background of a red tide by the dinoflagellate *Noctiluca scintillans* threatening to suffocate circular salmon farming pens in Southern Tasmania.

CHAPTER 20

# *Karenia brevis* and Florida Red Tides

## Major Problem for Tourism and Fisheries in Florida

Florida red-tides were first reported by Spanish explorers in the 1500s and the causative dinoflagellate initially called *Gymnodinium breve*. Florida Fish and Wildlife biologist the late Karen Steidinger first recognized that this species did not fit into *Gymnodinium*, and in 2000 Danish researchers renamed this species in honor of her, *Karenia brevis* [59]. During particularly large blooms there have been massive die offs in fish, seabirds, marine mammals such as manatees and dolphins and even turtles . The primary mechanism for such kills are thought to be brevetoxins. These toxins bind and open voltage gated sodium channels in cell membranes, which impacts on nerve cells. Most organisms cannot break down these toxins or excrete them, so they bioaccumulate in the marine food web. Although most invertebrates that consume *Karenia* cells directly are not impacted by this toxin, it is very toxic to animal or human consumers. The associated illness in humans is known as Neurotoxic Shellfish Poisoning (NSP) and includes a variety of neurological and gastrointestinal symptoms. and remnants of the toxin can be found throughout the food chain. This toxin also can impact humans as an aerosol because *Karenia* cells are easily broken during wave action. Inhaling aerosolized brevetoxin can lead to respiratory problems. Studies have shown that these brevetoxin aerosols can be blown from the coast as far as a mile inland during large blooms with strong onshore winds. Because of this, blooms are heavily monitored in these regions to detect brevetoxins in seafood and warn the public of aerosolized toxins along beaches. In 2025 in South Australia yellow slicks of related *Karenia cristata* blooms persisting for >15 months caused flu-like symptoms in surfers and beach visitors, but also killed fish, dolphins, sea horses, octopus and sharks. The costs to the local economy were estimated at AUD 250M. Local newspapers reacted with headlines of "*Algae Armageddon, Ocean-COVID, Sea Sick*". The unique ecological traits of this dinoflagellate to produce chemicals that suppress growth of competitors, feed on other algae, kill its predators and possibly even derive nutrients from died fish allowed it to develop into a persistent ecosystem disruptive algal bloom [60].

 | HTTPS://DOI.ORG/10.1515/9783112228265-020

CHAPTER 21

# *Gambierdiscus* and Ciguatera Poisoning

## Major Cause of Human Illnesses from Tropical Fish Consumption

Ciguatera Poisoning is a debilitating human illness, first reported in 1525 from Cameroon, West Africa, caused by eating tropical seafood contaminated by toxins which bioaccumulate from bottom-dwelling dinoflagellates. In 1774 Captain James Cook and his crew already suffered from the symptoms of ciguatera after consumption of red bass from the New Hebrides (Vanuatu). Ciguatera has been widely reported throughout the Caribbean ("*cigua*" is Spanish for "one poisoned from snails"), the Hawaiian Islands, French Polynesia, tropical Australia, Southeast Asia and the Indian Ocean. Recent reports are also available from the Canary Islands, Mediterranean, Gulf of Mexico, and the Atlantic off North Carolina [61]. Ciguatoxins are lipid-soluble and readily transferred through the food web from dinoflagellates, to herbivorous fish, to carnivorous fish, and ultimately to humans. Ciguatoxins are the most potent sodium channel toxins known. There is no immunity, toxins are cumulative, persistent, and in some 20% of patients neurological symptoms can persist for months or years. The disease often goes unreported or misdiagnosed. Acute symptoms appear within 48 hours after seafood consumption and include fatigue, pain, nausea, vomiting, diarrhea, low blood pressure, and heart rhythm disorder. The most diagnostic classic symptom is the sensation that cold things feel hot to the touch. Globally, ciguatera affects 50,000 to 500,000 people annually but fatalities are estimated to be less than 0.5%. The most common fish affected are groupers, barracudas, hogfish, snappers, carang, old wife, kingfish, parrotfish, Spanish mackerel and small surgeon fish (Fig. 14). Eating of these target fish species from known ciguatoxic areas is discouraged or sometimes forbidden, and the taking of only small fish is recommended because large fish may have accumulated too much toxin. Tracking down the origin of ciguatoxic fish in the market place poses a major challenge due to the mobility of fish. Ciguatera causes fish markets to ban product sales from entire estuaries (e. g., Platypus Bay, Australia), but impacts on human society penetrate far beyond the tropics through seafood trade, e. g., tropical fish import bans into European Union. In some instances, victims of ciguatera have sought financial compensation through legal action (e. g., against the Sydney fish market, Australia) adding significantly to the economic costs of any ciguatera event. Ocean warming and coral reef disturbance from coral bleaching and ocean acidification are predicted to increase ciguatera problems. The causative dinoflagellate *Gambierdiscus toxicus* was first identified in 1979 from the Gambier Islands in French Polynesia. However, as with other microalgae the use of molecular sequencing has now recognized at least 16 *Gambierdiscus* species of which only

 | HTTPS://DOI.ORG/10.1515/9783112228265-021

*G. australis*, *G. excentricus*, *G. pacificus*, *G. polynesiensis*, and *G. silvae* consistently produce ciguatoxins, while all *Gambierdiscus* species also produce maitotoxins but their role in the ciguatera syndrome is unclear.

FIGURE 14 Ciguatera fish poisoning information poster by the Florida Department of Health, against the background of a scanning electron micrograph of a *Gambierdiscus lapillus* dinoflagellate culture from the Australian Great Barrier Reef.

CHAPTER 22

# *Noctiluca* Night Lights

## Bioluminescence Attracts Tourism but Causes Problems for Submarines

South East Australian coastal waters are regularly lit up by bioluminescent spectacles caused by the plankton dinoflagellate *Noctiluca scintillans* ("scintillating night light", or "sea sparkle"). These single celled, bubble-shaped forms, 0.5 to 1 mm in diameter, carry a sticky tentacle to snare plankton prey that come too close. When they were first discovered in 1810 this caused them to be mistaken for small jellyfish. *Noctiluca* have no means of locomotion but can adjust their buoyancy to forage through the water column. When the food is finished, billions float to the surface in the hope of being carried by wind or currents to richer waters. If they end up washed ashore by onshore winds, the death throes of billions of stranded cells, stirred up by breaking waves create a spectacular light show (Fig. 15). These little fireworks have nothing to do with the chemical element phosphorus but use a molecule luciferin and an enzyme called luciferase, which combined with oxygen generates neon blue light. This biochemical strategy was first invented by the bacteria (Aristotle already described how damp wood can glow), but from there was borrowed by numerous other creatures to serve very different purposes such as for mushrooms to attract flies that spread spores, for glow worms to attract prey, and for fireflies to attract a mate. For sea creatures such as *Noctiluca*, the use of bioluminescence is thought to act as a burglar alarm to scare off predators. Ironically, almost everything we know about bioluminescence dates from Navy funded research during World War 1 after a stealth German submarine tried to sneak through the Strait of Gibraltar, but lit up like a Christmas tree and was sunk by Allied Forces. Despite all the research, these neon blue spectacles are difficult to forecast with precision. In daytime you can recognize *Noctiluca* as pink slicks but unlike other types of algal blooms this organism is essentially harmless to humans. The bloom responds to high diatom prey abundance during both summer and winter-spring, and when following calm water conditions the wind is blowing gently straight onto the beach, concentrating the bloom in one place. If it is on, grab your camera, don your wetsuit and get your children out of bed. You are in for an unforgettable treat. *Noctiluca* blooms have been increasing in the past 30–50 years in South East Australia, the North Sea, China and the Arabian Sea, but the reasons for this expansion are not yet fully understood [62].

 | HTTPS://DOI.ORG/10.1515/9783112228265-022

FIGURE 15 Top: Bioluminescence spectacle in Southern Tasmania. Bottom: Light micrograph of a Tasmanian bloom sample of the causative dinoflagellate *Noctiluca scintillans*.

CHAPTER 23

# Harmful Algal Blooms and Finfish Kills

## Threat to Global Seafood Security

Over 50% of the world's fish are currently provided by fish culture. On a global scale, the economic impacts of algal blooms on finfish aquaculture far exceed impacts on shellfish farms or tourism. Examples are the 1972 *Chattonella marina* flagellate bloom in the Seto Inland Sea, Japan (estimated USD 71 million loss to yellowtail aquaculture), the 1988 *Prymnesium polylepis* flagellate bloom in the European Kattegat with broad marine ecosystem impacts, and the 2015/16 *Pseudochattonella verruculosa* flagellate bloom in Chile which caused USD 800 million loss to salmon aquaculture. In addition to the large net pen operations in Norway and Chile, Scotland and Australia, small-scale pond fish farms predominate in Southeast Asia and India. The small golden-brown flagellate *Prymnesium parvum* first caused mass fish mortality of *Tilapia* brackish water culture ponds [63] in Israel in 1947 while in Texas fish kills by this "golden alga" have occurred since 1985 on the Pecos River, and in 2001 they destroyed a whole season's production of striped and hybrid striped bass. In Australia *Prymnesium* has caused mortality in Northern Territory barramundi ponds. Fish mortality can be caused by mechanical damage or clogging of fishes' gills, oxygen depletion during algal bloom collapse, but more commonly is caused by exudate chemicals (reactive oxygen radicals, free fatty acids) which irreversibly damage the gills of fish but typically are of no human health significance. Critical hurdles that limit progress in our understanding of ichthyotoxins and their control and mitigation include: Harmful algal blooms at fish farms are not usually a research priority until a major bloom occurs; data sharing between industry and scientists is very limited; and there is a lack of standardized methods to detect the causative "ichthyotoxins" in low concentrations dissolved in seawater [64]. Finfish aquaculture mass mortalities are considered economic disasters, and in most countries, the public, politicians and aquaculture industry expect some level of algal bloom control to mitigate their impacts. While wild fish can avoid and swim away from algal bloom, captive fish cannot, hence this is largely a human generated problem but, on the increase, caused by the need for seafood by an ever increasing human population. A widespread belief is that harmful algal blooms are increasing in frequency, intensity and global distribution. However, the first global scale statistical analyses of harmful algal bloom events in the period 1985–2018 revealed the lack of any uniform global trend in the number of harmful algal events once data were adjusted for regional variations in observational activities. Trends were instead variable and contrasting among different regions, which were characterized by different types of events, harmful species and emerging impacts. However,

 | HTTPS://DOI.ORG/10.1515/9783112228265-023

a clear relationship was found between the trends of harmful algal events and the monitoring effort associated with the development of aquaculture operations in coastal areas, leading to the conclusion that the latter is the source of the perceived increase in harmful algae events [65,66]. The UNEP Global Environmental Outlook lists the worldwide economic impacts of algal biotoxins on human health from seafood alone as USD 4 billion a year.

CHAPTER 24

# Zooxanthellae and Coral Bleaching

## Critical Component of Coral Reef Ecosystems under Threat of Climate Change

Coral reefs are amongst the most diverse ecosystems on Earth. When Charles Darwin first described coral reefs in 1842, he was puzzled how they could exist in such low nutrient environments with little to feed on. At the heart of all coral reefs lies the partnership between coral polyps, zooxanthella and bacteria. Zooxanthellae are single-celled dinoflagellates that live in symbiosis with marine invertebrates including sponges, corals, jellyfish, snails, clams, flat worms but also plankton radiolarians and foraminiferans. These dinoflagellates alternate between indiscriminate golden brown coccoid cells, and motile *Gymnodinium*-like dinoflagellate cells. When first discovered in the 1960s, a proposal was coined to rename coral reefs "dinoflagellate reefs" but this was never accepted by coral scientists. Most zooxanthellae belong in the genus *Symbiodinium*, and the species *Symbiodinium microadriaticum*, but these are now differentiated into eight genetic clades. One square meter of coral tissue can contain one to two million dinoflagellate cells. These photosynthetic algae pass on their photosynthetic products to their coral host animals in the form of sugars, glycerol and amino acids, and the algae in return gain carbon dioxide, phosphates, nitrogen compounds and shelter [67]. When corals are exposed to environmental stressors such as high temperatures (>2°C above summer average) this can result in the expulsion of zooxanthellae from host tissues. This strips the coral of its color, known as coral bleaching (Fig. 16). If short-lived then most bleached corals will recover, but if lasting >6 weeks most corals will not survive. Corals containing clade D zooxanthellae are more tolerant of higher temperatures. This may be used in engineering climate-proof coral strains in future [68]. Severe coral bleaching events simultaneously affecting multiple global coral reef systems occurred in 1998, 2010, 2014–2017 and 2020–2024. Globally coral reefs generate tourism worth USD 36 billion per year.

 | HTTPS://DOI.ORG/10.1515/9783112228265-024

FIGURE 16 Top: *Acropora* staghorn coral in the process of bleaching. Coral polyps on the outer branches have expelled their symbiotic algae, losing a source of food as well as their color. Bottom: Microscopic image of symbiotic zooxanthella dinoflagellates responsible for the entire productivity of coral reef ecosystems.

CHAPTER 25

# Green Algae: The Origin of Land Plants

Approximately 1.4 billion years ago the engulfment by a primitive colorless cell (believed to have been an Asgard Archaean; named after the home of the gods in Norse mythology) of green cyanobacteria led to the evolution of special organelles, called chloroplasts in plant cells. Similarly, the engulfment of purple sulfur bacteria led to the evolution of special organelles called mitochondria, used for respiration. Unlike the many different types and colors of plastids in different algal groups which were acquired on several occasions, the mitochondrion was acquired only once, explaining the broad similarity in metabolic processes in both plant and animal cells. The resulting cell types also contain a control center, called the *karyon* or nucleus. All our human body tissue is of this *eukaryotic* (with a true nucleus) cell type. This remarkable process of *serial endosymbiosis*, the engulfment of one organism by another, followed by their living together in partnership, was first advocated in 1960 by University of Chicago Lynn Margulis and has since been widely confirmed by the discovery of numerous examples in the present-day oceans. The most convincing evidence for this remarkable evolutionary process was provided in the 1990s by the discovery that mitochondria contain within their DNA molecular remnants of sulfur bacteria, and similarly plant chloroplast DNA contains molecular sequences that match cyanobacteria. As a result of symbiosis, microscopic plants contain all the structures necessary for life within the compass of a single cell. Early cyanobacteria gave rise to the various golden-brown algal groups (such as the diatoms, dinoflagellates and brown seaweeds) and eventually the green algae and green filamentous algae which were the precursors of the higher plants on the land. Symbioses are widespread among microbes (e. g., the ciliate protist *Paramecium bursarium* engulfing green *Chlorella* algae (Fig. 17). This process has been critical to evolution of life on our planet. It shows that life evolved mostly out of collaboration instead of competition as often thought. The green leaves of the trees on the land thus can be viewed as a "bloom" of microscopic green algal cells (the chloroplasts) living within the host body of a macroscopic plant. In the same way, the red blood cells in our human veins still resemble primitive single-celled animals floating in the original fluid of life, the salt solution of the sea. The red hemoglobin pigments of blood cells show a remarkable chemical similarity to the green chlorophyll pigments that color all green plant life on Earth. The grass green pigments of green algae include chlorophyll *b* that performs best under the high light spectral conditions on land and in shallow lakes and ponds. It is no coincidence that the human eye has its greatest sensitivity for green light (450 nm wavelength). Antarctic expeditioners grow green plants in glass houses not so much to eat as food but because it gives them comfort to survive the long dark winter.

 | HTTPS://DOI.ORG/10.1515/9783112228265-025

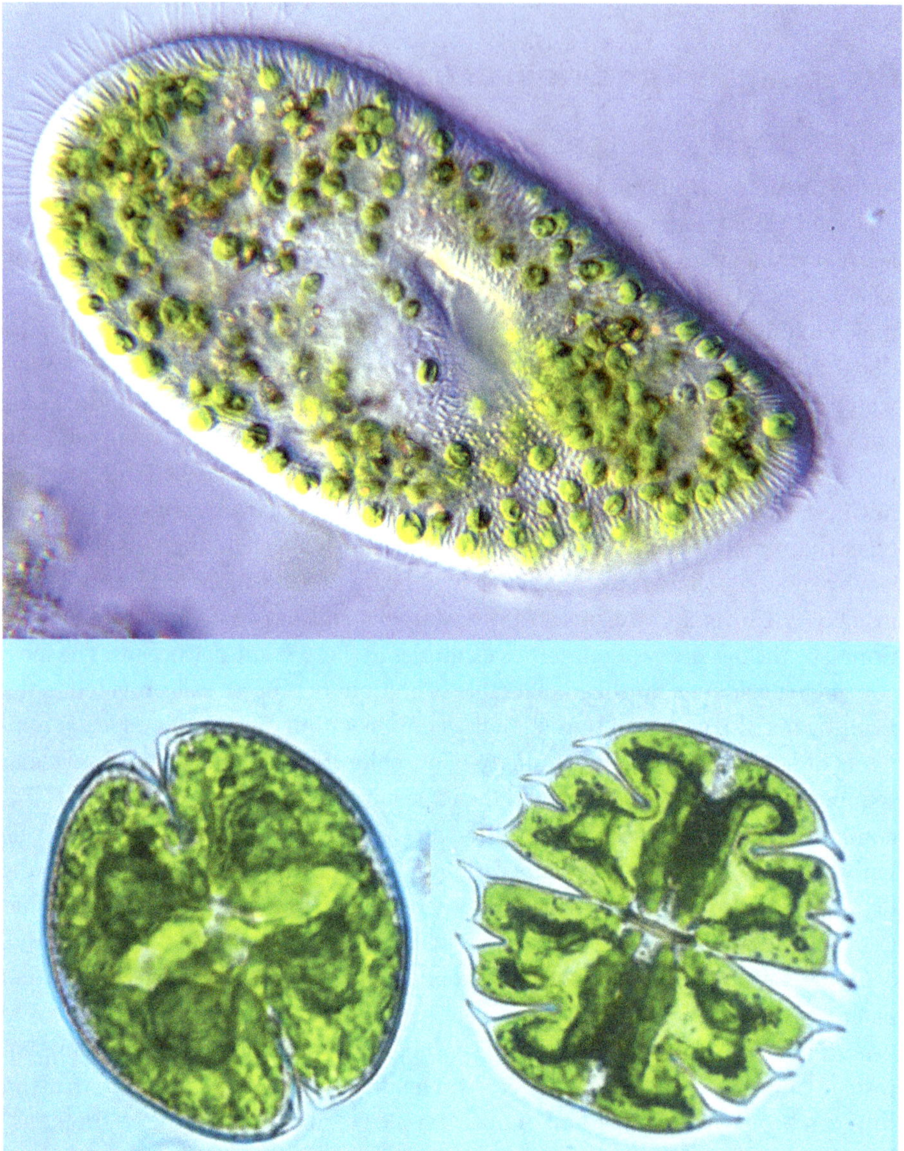

FIGURE 17 Top: The protozoan ciliate *Paramecium bursarium* which feeds on coccoid green *Chlorella* type algae but which it keeps alive as in a green house. This is how the green chloroplasts of land plants originated. Bottom: Light micrographs of two species of the colonial freshwater green alga *Cosmarium*.

CHAPTER 26

# *Dunaliella salina* and β-carotene

## Important Algal Biotechnology Industry

The green alga *Dunaliella salina* was named after its discoverer Michel Felix Dunal, who first reported in 1838 sighting the organism in saltern evaporation ponds in Montpellier, France. Under high light and high salinity conditions, this normally green alga turns red to protect the organism from long-term UV radiation by producing up to 5% of its dry weight as β-carotene. This carotenoid pigment (first purified from carrots by Swiss scientist Paul Karrer in 1931) is a precursor to vitamin A and widely used as a food supplement to create yellow butter or orange fruit juice. From a first commercial pilot plant established in the USSR in 1966, the commercial cultivation of *Dunaliella salina* has been pursued in Australia (since 1980), Israel and the USA. Grown in 5 ha open air ponds, companies like Western Biotechnology in Australia (Fig. 18) produce USD 5M annually (market price is $600/kg [69,70]). Similar ventures in the US focus on astaxanthin carotenoid production from the green alga *Haematococcus pluvialis* (valued at USD 2,000–$3,000/kg) used as a supplement in feed pellets for salmon aquaculture to maintain their pink flesh. Similarly, flamingos in zoos would lose their pink plumage without carotenoid dietary supplements. Carotenoids cannot be synthesized by animals and therefore must be obtained from their algal or plant diet. The global carotenoid industry currently is valued at USD 1.5 billion per year [71]. Blood-like red snow by the related green alga *Sanguinea nivaloides* similarly derives from excess carotenoid production. Charles Darwin himself came across white snow in the Andes Mountains that turned red as it thawed. Bulk algal cultures of *Dunaliella*, grown in protective 3D printed translucent bioplastic beakers, have even been suggested as a means to construct extraterrestrial human habitats such as on Mars where ice and carbon dioxide are found [72].

 | HTTPS://DOI.ORG/10.1515/9783112228265-026

CHAPTER 27

# *Botryococcus*: Good and Bad Oils

*Botryococcus* is a globally distributed freshwater green alga that produces spectacular green or yellow scums. It synthesizes abundant unique hydrocarbons (25–70% of dry weight) which can be squeezed out by moderate pressure without destroying the cells. This organism is the source of boghead coal and torbanite deposits in Scotland that are used in the paraffin industry. Similarly, the alga produced black rubbery and highly flammable coorongite deposits in South Australia. Accordingly, major efforts have been made to mass culture this and other species (*Nannochloropsis, Chlorella, Scenedesmus*) for biodiesel production. Microalgae with high oil content can be grown on land in open pond systems or more expensive to operate photobioreactors. So far slow growth rates [73] and the high cost of oil refinery from watery extracts (triple that of fossil fuel) have hindered progress [74].

Human populations that eat significant amounts of seafood such as in Japan and the Mediterranean tend to be healthier and live longer. This has been largely contributed to the presence in seafood of omega-3 long-chain polyunsaturated fatty acids (PUFAs) which can only be synthetized by microalgae but not land plants [75]. Because growing bulk cultures of microalgae never can supply human needs for "good oils", scientists cleverly identified the responsible algal genes and successfully inserted them in *Canola* mustard seed agricultural crops. Ultimately, it will be a societal decision whether we accept these genetically modified crops but then we may not have a choice in future the way humankind is going. Some microalgae can produce up to 20 times more proteins per ha per year than soybean plants, and represent an alternative to environmentally costly meat production while having a much more limited climate footprint, with no deforestation nor methane emissions.

 | HTTPS://DOI.ORG/10.1515/9783112228265-027

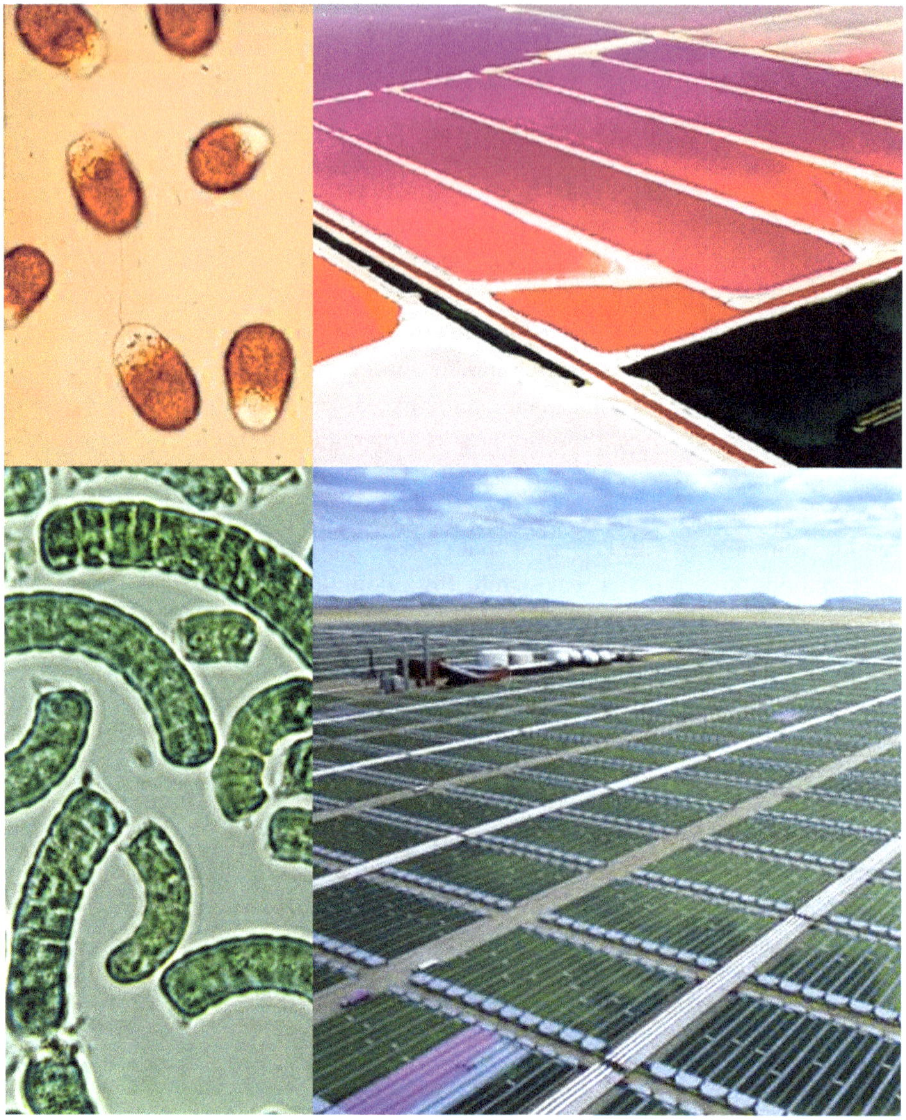

FIGURE 18 Examples of algal biotechnology. Top: Betatene pond culture of the green alga *Dunaliella salina* in Western Australia used for the production of β-carotene. Bottom: Pond cultures of the cyanobacterium *Spirulina* for biofuel production in the Californian desert.

# PART 3

# Protozoa

CHAPTER 28

# Radiolarians: Art Forms in Nature

Microscopic plankton cells have fascinated scientists like Ernest Haeckel and artists such as Joan Miro, Paul Klee and Vasily Kandinsky, ever since they were first discovered. The appeal of these architecturally interesting and aesthetically beautiful organisms to a large extent stems from their symmetrical design. When we humans reduce a complex pattern to a repetition of some elementary design, this brings with it a simplification and visual harmony. Repetitive patterns have been common through the ages and in many cultures, in textile design, decoration and architecture. Architects and microscopic plankton are both faced with the challenge of how to manufacture the strongest possible structure using the least amount of building material. The striking similarity between the design of the diatom cell wall of *Pyxidicula* and the geodesic dome patented by Buckminster Fuller (pictured in the Montreal 1967 Expo building) is the result (Fig. 19).

Aesthetically fascinating and almost mystically beautiful as these miniature organisms may appear to the human mind, one needs to avoid reading too much in the purpose of their design. Symmetry in these primitive life forms is a consequence of the manner in which their cell walls are synthesized from a unit casting mold. In the coccolithophorids, minute body scales are formed inside the cells in special vesicles before being transported to the external surface of the cell where they are deposited and function as a template for calcium carbonate crystal growth. The formation of the ornamented diatom cell wall is orchestrated by dedicated silicon deposition vesicles underneath the outer cell membrane. In here, minute templates act as scaffolds for the erection of rigid geometric structures. A detailed understanding of this manufacturing process is of considerable interest in nanotechnology.

Initially Radiolaria were assumed to be multicellular animals until Haeckel coined the new kingdom Protista in 1866. Radiolaria possess delicately sculpted skeletons of silica or in the case of the subgroup *Acantharia* of strontium sulphate. Ernest Haeckel [76] working at the University of Jena, Germany, published his first monograph on these little known microscopic cells in 1862 at age 28, and went on to describe 144 new species and 47 genera (Fig. 20). In 1899 he embarked on the publication of a series of 100 splendid plates *Künstformen der Natur* to make accessible to the educated general public "those marvelous treasures of beauty that lie hidden in the depths of the ocean". To honor Haeckel, in 1900 the French architect René Binet erected a Porte Monumentale at the Place de la Concorde in Paris which served as the main entrance to the world Expo. This portal was modelled on the radiolarian species *Clathrocanium reginae* and studded with alternating colors symbolically transporting visitors back to the beginnings of evolution.

 | HTTPS://DOI.ORG/10.1515/9783112228265-028

FIGURE 19 Architects using protists for inspiration. Top: Buckminster Fuller's geodesic dome structure as used for the Montreal 1967 Expo building bears a striking resemblance to the diatom cell wall of *Pyxidicula* (resting spore of *Stephanopyxis*). Middle: The leaning tower of Pisa resembles the chain forming diatom *Paralia sulcata*; Bottom: René Binet's Porte Monumentale at the Place de la Concorde in Paris which served as the main entrance to the 1900 world Expo. This portal was modelled on the radiolarian *Clathrocanium reginae*. (see Fig. 20).

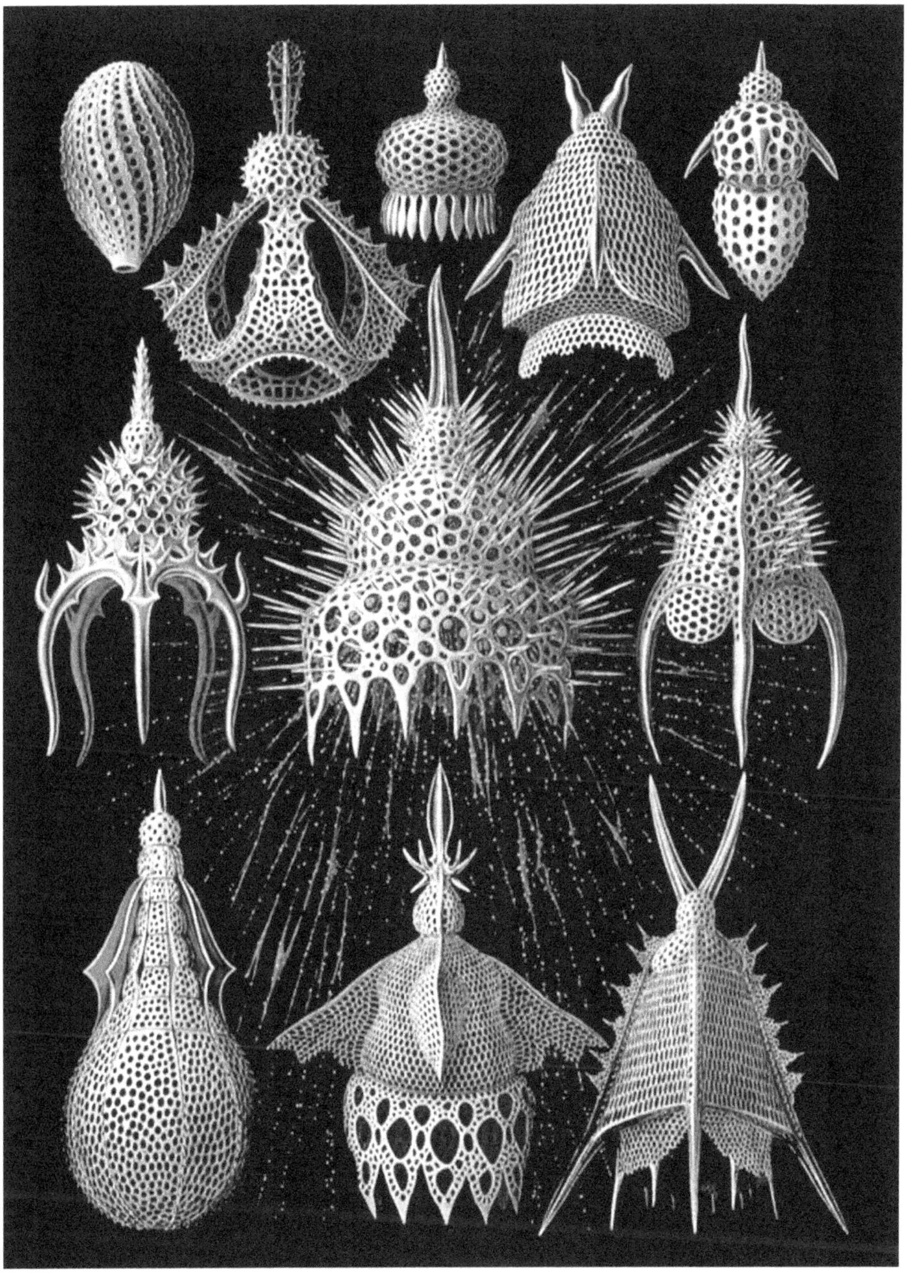

FIGURE 20 Radiolaria. Plate 31 from Haeckel, E. 1899–1904. *Künstformen der Natur*.

CHAPTER 29

# Tintinnid Loricas: Master Home Decorators

Another group of single celled protozoa, the tintinnids, live in a protective flask-like house called lorica. They feed themselves in watery habitats by creating feeding currents by beating hair-like rows of cilia. Some of these species discard feeding waste products from their prey, notably calcareous coccolithophorid scales and sometimes diatom fragments, to decorate their houses (Fig. 21 [77,78]). Doing so, these tiny animals can add up to 25% of ballast to their lorica weight, allowing them to adjust their depth position in the water column, but also adjust swimming speeds to enhance further food capture. Tintinnids are not selective but pick what is available at the time, except for rejecting larger items. This behavior is not unlike human dwellings which inadvertently may show off their inhabitants' feeding and drinking habits via the contents of their rubbish and recycling bins filled with oyster shells, empty wine and beer bottles. Humans can learn from microbes to become more creative in utilizing and recycling waste to decorate and build homes and gardens!

 | HTTPS://DOI.ORG/10.1515/9783112228265-029

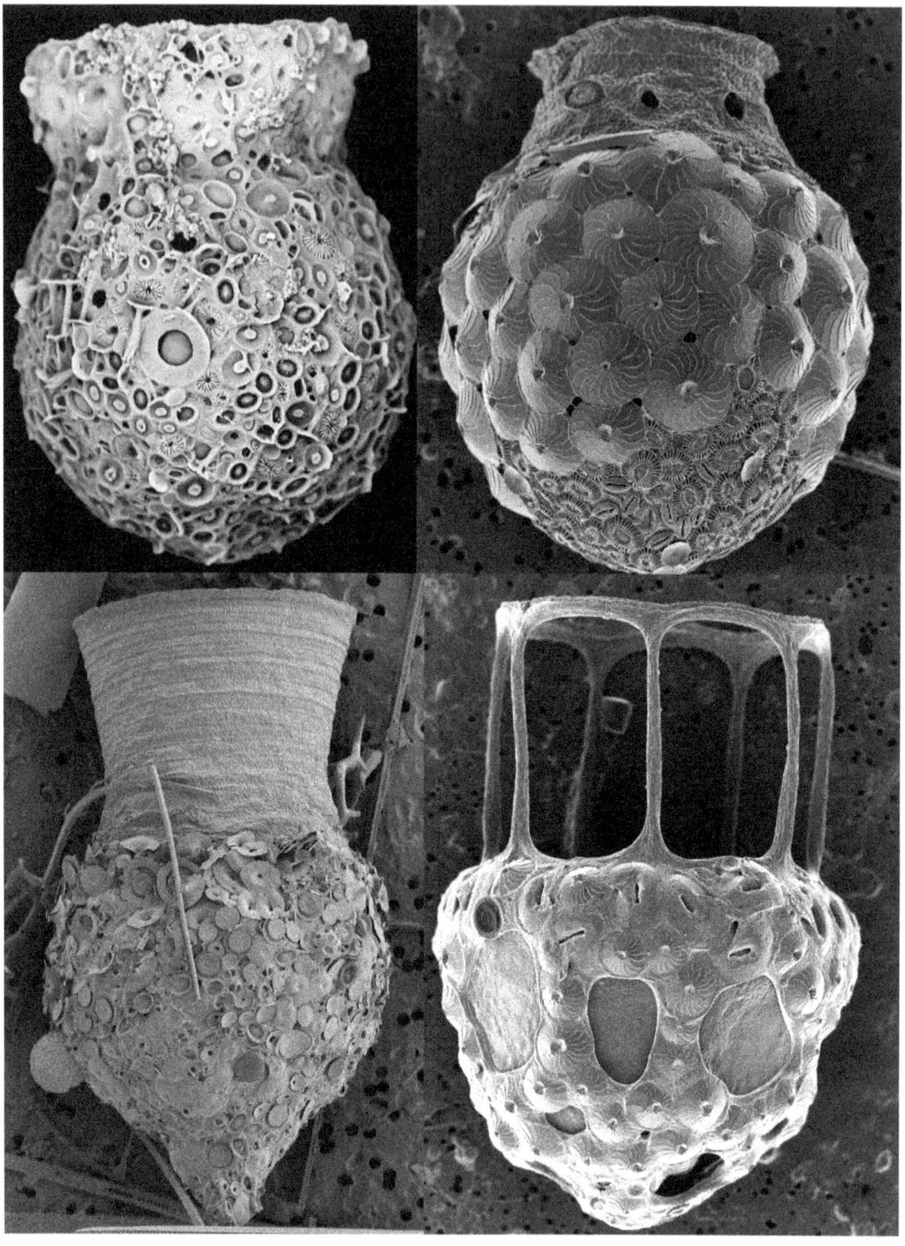

FIGURE 21 Scanning electron micrographs of planktonic tintinnid loricae decorated with coccolithophorids from the surrounding seawater. Top left: *Codonella galea* from tropical Tonga waters; Top right: *Codonellopsis pusilla* from the Southern Ocean; Bottom left: *Codonellopsis pura* from Coral Sea; Bottom right: *Dictyocysta lepida* from Southern Ocean.

CHAPTER 30

# *Plasmodium* and Malaria

## Most Infectious Human Tropical Disease that has Influenced Evolution of the Human Immune System

Malaria, a swamp fever syndrome associated with bad air (*mal' aria*), was first linked to *Plasmodium* parasites in *Anopheles* mosquitoes by Ronald Ross in 1902. To date we recognize five *Plasmodium* protozoan species that cause malaria in humans. The deadliest *Plasmodium falciparum* is prevalent on the African continent while *Plasmodium vivax* is dominant in countries outside sub-Saharan Africa. Other *Plasmodium* species can also infect birds, lizards, rodents and other primates. The highly virulent *Plasmodium falciparum* increased in abundance around 10,000 years ago coincident with the beginnings of human agricultural settlements. This protozoan is thought to have originated from a *Plasmodium* species in African gorillas. Strikingly, native sub-Saharan human populations are currently completely resistant to malaria, whereas all other ethnic human populations are vulnerable [79]. Unravelling the molecular basis for such disease resistance is being explored toward a new vaccine under trial. Malaria parasites harbor a relict plastid, homologous to the chloroplasts of plants and algae, which may be another target in the search for anti-malarial drugs [80].

*Plasmodium* is spread to humans through the bites of infected mosquitoes. The parasite enters red blood cells but without destroying them (Fig. 22). People who have malaria suffer from high fever, sweats, shaking chills, headaches, diarrhea and fatigue. During the second half of the 20th century, efforts to reduce mosquito breeding grounds by draining wetlands and spraying standing water with the insecticide DDT sharply reduced malaria in high income countries such as the United States and Europe. However, the disease is still common today in tropical and subtropical countries. Each year 300 to 500 million people are infected and an estimated 2.7 million per year die of the disease. Quinine and chloroquine extracted from the tree bark of *Cinchona* to this day remain an effective treatment, together with artemisinin from *Artemisia* sweet wormwood. Protective clothing, bed nets and insecticides can protect you while traveling. One can take preventive medicine before, during and after a trip to a high-risk area, but unfortunately many malaria parasites have developed resistance to common drugs.

*Not to be confused*: Yellow fever which is transmitted by infected *Aedes* and *Haemagogus* mosquitoes is instead caused by a virus and has very different epidemiological dynamics. The latter disease is more common in the Caribbean and can only be treated by a vaccine and does not respond to quinine or artemisinin.

 | HTTPS://DOI.ORG/10.1515/9783112228265-030

FIGURE 22 Top: Mosquito *Aedes aegypti* sucking human blood. Bottom: *Plasmodium* protozoan parasite (green) among human red blood cells.

# PART 4

# Fungi

# Introduction to Part 4

Fungi are best known as mushrooms, moulds, mildews and single-celled yeasts. They are critical in the nutrient cycling of soils, i. e., the forest mycelium depicted in James Cameron movie *Avatar*. They can also cause diseases such as Dutch elm disease, athlete's foot, but at the same time are highly beneficial to humans in the manufacture of yoghurt, beer, cheese and antibiotics such as penicillin and streptomycin. Historically fungi were treated as plants, but they do not photosynthesize and instead extract food from decaying organic material (called saprophytic) while their chitin cell walls more resemble the shells of insects or claws of mammals and are not at all like those of land plants.

 | HTTPS://DOI.ORG/10.1515/9783112228265-031

CHAPTER 31

# Baker's Yeast *Saccharomyces*

## Indispensable for Human Society to Make Bread, Beer and Wine

The word yeast comes from Old English *gyst* and the Indo-European *yes*, meaning "bubble". The name yeast is often used for the fungal species *Saccharomyces cerevisiae* but in fact reflects a broad range of organisms. These cells divide by budding (Fig. 23) and are one of the earliest domesticated microbes used to make bread, beer and wine. Archaeologists have retrieved yeasts in Israel dating back 5,000 years ago, from vessels believed to have contained alcoholic beverages such as beer and mead. Anton van Leeuwenhoek first observed yeast under a microscope in 1680, but he did not consider them to be living organisms. Only in 1837 did German cell biologist Theodor Schwann recognize them to belong to what are now called Fungi. French microbiologist Louis Pasteur first proved that alcoholic fermentation was conducted by living yeasts and not by a chemical catalyst. The alcohol is produced by the yeast feeding on sugar in the absence of oxygen [81]. In beer making, the sugar comes from germinating grains of barley, while wines are produced by the species *Saccharomyces ellipsoideus* naturally found on the skin of grapes. In bread making, the yeast does not produce alcohol but carbon dioxide gas which causes the dough to rise. In 1996 the complete genome of *Saccharomyces cerevisiae* was sequenced. This allowed Danish biotech company Novo Nordisk to metabolically engineer this ancient "microbial workhorse" to produce insulin for the treatment of diabetes.

 | HTTPS://DOI.ORG/10.1515/9783112228265-032

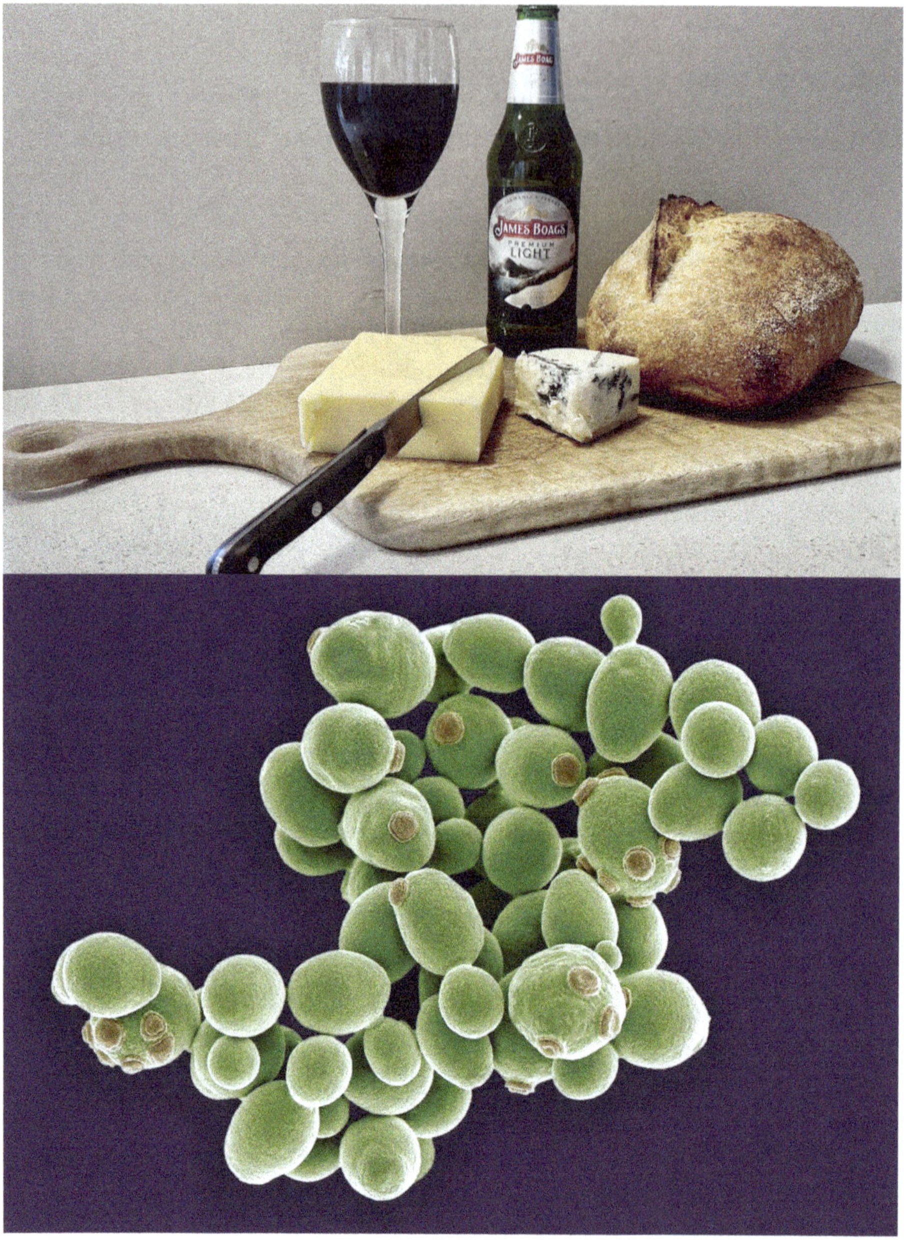


FIGURE 23 Yeast is one of the earliest domesticated microbes, dating back 5,000 years ago. Without yeast fungus, we would not be able to enjoy bread or wine or beer. Without *Penicillium* fungus we would not have French cheese.

CHAPTER 32

# *Penicillium*: The Birth of Antibiotics

The idea that mould could be a cure for the treatment of wounds and disease dates back to Greek and Roman folk medicine. French microbiologist Louis Pasteur first reported that airborne organisms could stop the growth of *Anthrax* bacteria. However, it wasn't until 1928 when Alexander Fleming returned from holiday, that he noted that contamination by the fungus *Penicillium notatum* prevented his golden staph *Staphylococcus aureus* bacterial cultures from growing. He found that the fungal secretion could kill many Gram-positive bacteria. The purification of the causative chemical substance *penicillin* (Fig. 24), spearheaded by a team of British scientists led by Howard Florey and Ernst Chain, is one of the success stories in the history of medicine [81]. Initially gallons of broth were required to produce a fingernail of penicillin, but eventually its mass production was achieved in special fermentation vessels. During World War II, penicillin made a major difference in the number of deaths and amputations among allied forces with infected wounds. Penicillin causes cell walls of many species of bacteria to disintegrate during cell division, and therefore the bacteria cannot replicate and die. To date commercial antibiotic production of penicillin is based on *Penicillium chrysogenum*, while *Streptomyces griseus* is being used to produce streptomycin as more widely applied against Gram-negative bacteria. Antibiotics are now a USD 25 billion per year business. Unfortunately, their present day overprescription has resulted in many antibiotic resistant bacterial strains causing a looming global health crisis by rendering critical antimicrobial drugs to become ineffective. In 2021 some 4.7 million human deaths were associated with Anti Microbial Resistance (AMR [82]). Other *Penicillium* species such as *Penicillium camembertii* and *Penicillium roqueforti* are beneficial to humans in the manufacture of cheeses such as Camembert and Roquefort. Another fungal species *Aspergillus niger* is widely used in biotechnology for the production of enzymes, such as pectinase for clarifying wine and fruit juice, glucoamylase for the production of glucose, and protease for cheese production.

*Not to be confused*: Antibiotics do not work against viruses.

 | HTTPS://DOI.ORG/10.1515/9783112228265-033

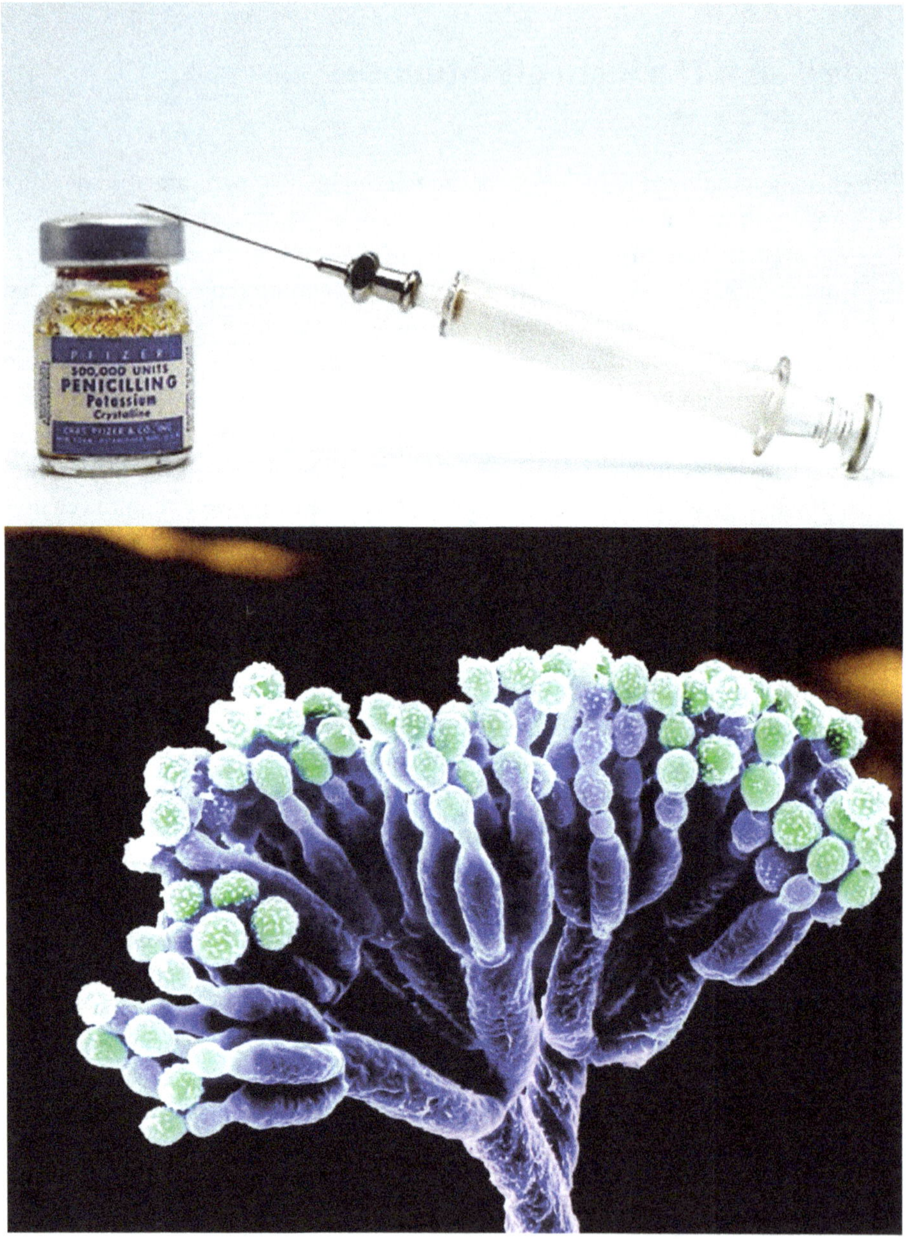


FIGURE 24 Commercial penicillin product. Below: Scanning micrograph of *Penicillium* fungal spores.

CHAPTER 33

# Soil Fungi and the Wood Wide Web

## Critical for Nutrient Cycling in Soils

For a long time, fungi were thought to be the enemy of plants and trees, spreading disease and consuming the trees they killed. However, over the last 30 years a greater appreciation has emerged of the positive role that fungi play in the lives of trees and plants. *Chaetomium globosum* is one of the most prevalent fungi on our planet of vital importance for the recycling of decaying plant material. This fungus can also be common in damp buildings in North America and Europe causing wood rot. It can decay a variety of wood types such as aspen and pine by releasing enzymes from its filamentous hyphae that digest plant fibers into smaller sugar molecules [83]. Other fungi such as *Glomus* produce a threadlike network of hyphae – called mycelium – around or inside plant roots. This and related fungal species attracted considerable attention in a 1997 study by Suzanne Simard [84–86] and collaborators from the University of British Columbia who showed that 6% of radioactively labelled carbon could be passed from birch to fir trees that shared an underground mycelium network. No such communication was observed between birch and cedar that do not have such network. The mycelium interacts with the roots of trees and plants in a symbiotic relationship whereby the tree provides the products from photosynthesis to the fungus, since the fungus doesn't have green leaves. As the mycelium grows through the soil it picks up nutrients and water and then transports them back to the tree. The concept of a "wood wide web" of communication and of sharing of resources between plants via the fungal mycelium was made famous by the movie *Avatar* (Fig. 25). However its quantitative importance for forest management and the chemical nature of the signal molecules remain to be fully elucidated [87]. This recognition of the key role of soil microbes has highlighted the ill-conceived detrimental practice of clear-felling forests or ploughing agricultural fields for the purpose of creating monocultural crops.

 | HTTPS://DOI.ORG/10.1515/9783112228265-034

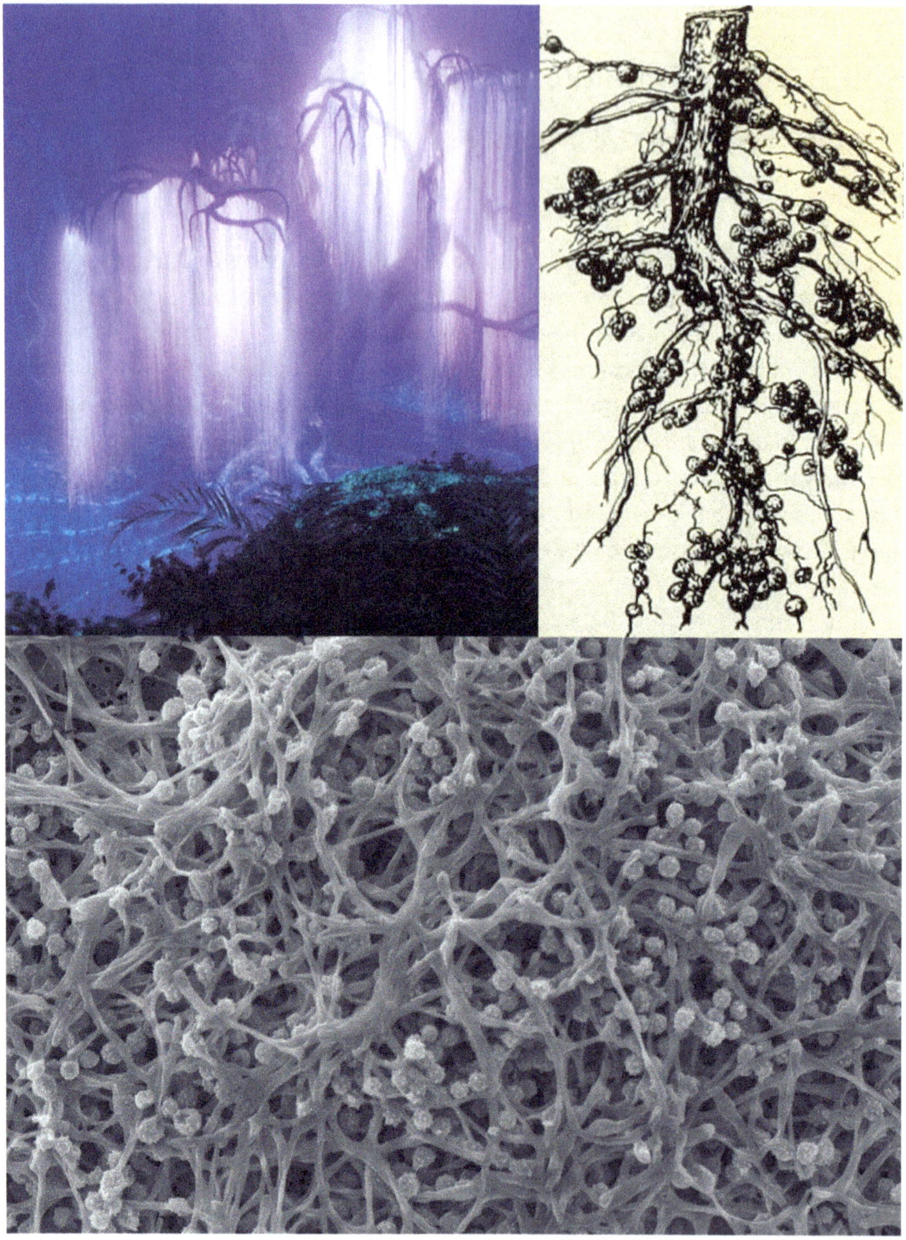

FIGURE 25 Top left. The forest floor microbiome comprises a fungal mycelium as visualized in the movie *Avatar* "Tree of Souls". Top right: nitrogen fixing bacteria such as *Rhizobium* in root nodules. Bottom: Fungal mycelium and fungal spores of *Aspergillus sydowii* from an Australian dust storm event.

CHAPTER 34

# Fungal Toxins in Food

## Global Threat to Agricultural Food Security

Fungal toxins or mycotoxins generally only affect humans with underlying health problems or a weakened immune system such as caused by chronic lung disease, tuberculosis, HIV, cancer and diabetes, or critically ill patients in hospital care. But there are exceptions. In Mediaeval times rye and related cereals contaminated with the fungus *Claviceps purpurea* caused people to suffer from ergotism or Saint Anthony's fire. The symptoms included burning sensations in the limbs as well as hallucinations. It was not until the 1930s that Swiss chemist Alfred Hoffmann extracted the alkaloid compound LSD (lysergic acid diethylamide) which became a popular recreational hallucinogenic drug in the 1960s. Aflatoxins are produced by certain moulds such as *Aspergillus flavus* and *Aspergillus parasiticus*, which grow in soil, decaying vegetation, hay and grains. Crops that can be contaminated include cereals, oilseeds, spices and tree nuts, but the toxins can also be found in the milk of animals fed with contaminated feed. Large doses of aflatoxins can lead to acute poisoning and be life threatening, usually through damage to the liver, but also have been shown to damage human DNA and cause cancer. Ochratoxin A is another fungal toxin produced by several species of *Aspergillus* and *Penicillium*. Contamination of cereals and cereal products, coffee beans, dry vine fruits, wine and grape juice, spices and licorice occurs worldwide. Ochratoxin A is known to cause kidney damage but may also affect fetal development and the immune system. Patulin is produced by *Aspergillus*, *Penicillium* and *Byssochlamys*. Often found in rotting apple products (including apple juice), patulin can also occur in various mouldy fruits, grains and other foods. Acute symptoms in animals include liver, spleen and kidney damage and toxicity to the immune system. In humans, nausea, gastrointestinal disturbances and vomiting have been reported, but patulin can also damage DNA. *Fusarium* fungi are common in soil and produce a range of different toxins. Different *Fusarium* toxins are associated with certain types of cereal. For example, both deoxynivalenol and zearalenone are often associated with wheat, T-2 and HT-2 toxins with oats, and fumonisins with maize. Trichothecenes can be acutely toxic to humans, causing rapid irritation to the skin or intestinal mucosa and lead to diarrhea. Chronic effects in animals include suppression of the immune system. Zearalenone has been shown to have hormonal, estrogenic effects and can cause infertility particularly in pigs. Fumonisins have been related to esophageal cancer in humans, and liver and kidney toxicity in animals. People who have enough to eat can avoid foods contaminated by moulds, hence exposure to acute levels of fungal toxins is rare in developed countries. Problems are exacerbated whenever shipping,

 | HTTPS://DOI.ORG/10.1515/9783112228265-035

handling, and storage practices are conducive to mould growth. The economic consequences of mycotoxin contamination are profound, and it is important to have policies in place that ensure that "hidden" fungal toxins do not pose a significant hazard to human health [88]. The "curse" of illnesses and deaths of scientists who entered the ancient tomb of King Tutankhamun in Egypt has been circumstantially linked to reawakening harmful *Aspergillus* fungi after lying dormant for thousands of years.

CHAPTER 35

# Fungal Pathogens Impacting Viticulture

The cultivated grapevine *Vitis vinifera* first evolved in Europe and has no genetic resistance to fungal pathogens from North America. Mildew is a thin whitish coating of surface growth of fungal hyphae on living plants. Powdery mildew is disease caused by the fungus *Erisyphe necator* which affects a wide range of plants, including apples, peas and roses. In grapevines, it causes reduced yield and loss of berry and wine quality. A closely related pathogen for viticulture is the fungus *Botrytis cinerea* or grey mould. By contrast, downy mildew *Plasmopora viticola* is an oomycete pathogen (not a fungus). It can be a problem for grapes and vegetables such as cucumber and pumpkin and affects both leaves and berries. Downy mildew requires a rain event for infection but when uncontrolled can cause defoliation and complete crop loss.

Wine growers typically cultivate more vulnerable rose bushes in their vineyard, not for decoration but to serve as early warning of fungal diseases to trigger fungicide applications on their vines. In 2002, grapevines accounted for 6% per cent of the total crop area in Europe, while 70% of all fungicides used were applied to grapevines.

 | HTTPS://DOI.ORG/10.1515/9783112228265-036

CHAPTER 36

# *Aspergillus*, Coral Disease and Dust Storms

While fungi are best studied in terrestrial environments, they also occur widespread in marine environments as decomposers, parasites and pathogens. Dust storms in the Caribbean, Africa and Australia are a well-known mechanism to transfer fungal spores into marine environments. When in 1832 Charles Darwin sailed past the Cape de Verde archipelago he collected Sahara dust from the masthead of his ship which included fungal and bacterial spores that were still viable almost 200 years later [89]. Salt tolerant species such as *Aspergillus sydowii* occur equally in both terrestrial and marine habitats. In 1997 and 2003 this fungus was associated with an outbreak of "aspergillosis" disease causing 20 to 90% mortality of coral fans *Gorgonia ventalina* in the Caribbean. While pathogenic and non-pathogenic fungal strains were genetically similar, they produced slightly different metabolites [90]. These fungal chemicals (sydowic acid, sydowinol, sydowinin) adversely impact on *Symbiodinium* coral symbionts, in a manner somewhat similar to coral bleaching [91,92]. This suggests that corals under stress from nutrient pollution, land runoff, dust storms and climate change are more vulnerable to opportunistic fungal pathogens. Other *Aspergillus* species, such as *Aspergillus sojae* and *Aspergillus oryzae* are instead important in the fermentation of soybeans and wheat to produce soy sauce.

 | HTTPS://DOI.ORG/10.1515/9783112228265-037

# PART 5

# Bacteria

## Introduction to Part 5

Bacteria range in size from 0.3 to 20 µm and come in three basic shapes: spheres (cocci), rods (bacilli) or spirals (spiralla). They may also associate in groups, e. g., streptococci. Historically, bacteria were grouped into two main categories: Gram-positive bacteria with a thick wall, and Gram-negative bacteria with a thin layer of peptidoglycan within the cell wall. In 1853 Danish microbiologist Hans Christian Gram designed a staining protocol using crystal violet/ safranin to discriminate between them. However, recent molecular sequencing has led to the further recognition of twelve bacterial groups, with only two of them Gram-positive and the other ten categories further divided based on the content of their genetic materials, the DNA nucleotides guanine and cytosine. Bacteria are much feared as the cause of human diseases such as the pest, cholera, smallpox, but most are beneficial for humankind.

 | HTTPS://DOI.ORG/10.1515/9783112228265-038

CHAPTER 37

# Human Microbiome: You Are What You Eat

## Key Driver of Human Health

The human microbiome consists of a community of protozoa, archaea, fungi, viruses and predominantly bacteria that live symbiotically on and within various sites of the human body, especially the gut. The microbiome is estimated to tally around $10^{13}$ to $10^{14}$ cells, that is a 1:1 microbe to human cells ratio and comprises 300 to 1,000 species with an estimated staggering total weight of one to 1.5 kilogram. As a community they harbor 50 to 100-fold more genes than the host does. Relatively few species occur in the highly acidic stomach and small intestines, while the colon contains the highest densities of microbes. Bacteria also make up 60% of the mass of dry feces. The incidence of the bacterium *Escherichia coli* in recreational swimming waters is widely used to monitor for human fecal sewage pollution. At the same time, systematic monitoring of human sewage plants could also warn of future new virus outbreaks, as was first explored during COVID. Travelling to foreign countries exposes humans to bacteria in drinking water with which our bodies are not familiar, resulting in diarrhea. The causative problem bacteria belong to *Campylobacter, Escherichia coli, Salmonella, Shigella, Yersinia* and *Vibrio*. While most *Escherichia coli* variants are harmless, a few can cause human diseases such as diarrhea, sepsis and urinary tract infections. By contrast *Shigella*, which is a close relative, can kill humans after ingesting just ten individuals. Human gut microbiota are established one to two years after birth and acquired through parent-to-child transmission and transfer from food, water and other environmental sources. Caesarean birth delivery [93] influences the composition of the neonatal gut microbiota, with higher abundance of harmful bacterium *Clostridium difficile* in Caesarean infants increasing the risk of certain infections, while the beneficial *Bifidobacterium* is more prevalent in vaginally delivered infants which exhibit improved health. Probiotics provide us with live beneficial bacteria, including *Lactobacillus* (e. g., *L. casei, L. delbrueckii bulgaricum, L. rhamnosus*) and *Bifidobacterium*, often delivered in yoghurt-like drinks, to improve gut health. Our gut flora changes with age, geography, diet and overall health. Gut microbes play a role in defense against pathogens, metabolism of indigestible food, while increasing evidence points to their role in training the immune system and biochemical signaling between the gastrointestinal tract and the central nervous system (referred to as the “gut-brain axis”). Extensive changes in gut microbiota can be associated with obesity, while other species can cause disease (e. g., inflammatory bowel disease) or increased cancer risk (e. g., colon cancer). In 1984 Australian professors Barry Marshall and Robin Warren discovered that

 | HTTPS://DOI.ORG/10.1515/9783112228265-039

the bacterium *Helicobacter pylori* can initiate formation of stomach ulcers previously associated with mental stress. Marshall famously drank the bacteria himself to prove it caused stomach inflammation which then could be cured by a course of antibiotics to kill the bacteria. Allergy and asthma have also been linked to gut flora imbalance [94]. Other diseases such as celiac disease, and autism spectrum disorder may also have a relation with our gut microbiome. Microbiomes of people suffering from depression tend to lack *Coprococcus* and *Dialister* bacteria. On the other hand, *Roseburia intestinalis*, *Ruminococcus gnavus*, and *Faecalibacterium prausnitzii* are indicators of good gut health. In response to this rapidly increasing body of new knowledge, novel therapies are now being developed such as fecal microbial transplantation, prebiotics (health food for microbes, such as dietary fiber), probiotics (live microbes) and postbiotics (fermented foods such as sauerkraut, kombucha) to target associated diseases. Idioms such as "having a gut feeling about something" or "you are what you eat" literally have taken on a whole new meaning [95]. While home test kits for our gut microbiome are on the horizon, at present the science behind fecal DNA analysis is still limited by methodological problems and lack of comprehensive genetic reference data bases.

*Not to be confused*: Gut microbiomes also include fungi such as *Candida* and *Cladosporium*.

CHAPTER 38

# *Yersinia pestis* and the Black Death

## Responsible for More Than 50 Million Deaths in Medieval Europe

The bacterium *Yersinia pestis* was discovered in 1894 by Alexandre Yersin, a Swiss French physician and bacteriologist from the Pasteur Institute, during an epidemic of the plague in Hong Kong [96]. Earliest Chinese descriptions of the Bubonic Plague date back to 610 AD. The plague (from the Latin *plaga* = to strike or wound) arrived in Europe in Constantinople on grain ships from Egypt infested with rats, from an origin either in northeastern India or central Africa. Ancient DNA samples from the dental pulp of buried victims retrospectively confirmed that *Y. pestis* was also the cause of the Plague of Justinian in the 6th to 8th century, the Medieval Black Death in the 14th century, and the Great Plague of London from 1665 to 1666. The term "Black Death" seems to have originated from a mistranslation of the Latin *atra mors*, in which *atra* can mean both "horrible" and "black". In the 14th century the Mongols catapulted corpses which had died of the plague into the city walls of Caffa (now Ethiopia) as an early form of biological warfare. The most common form of plague (Bubonic Plague) results in swollen and tender lymph nodes in the groin, armpits or neck called "buboes". The rarest but deadliest form "Pneumonic Plague" affects the lungs and can be spread from person to person. Bubonic Plague has a fatality of 30 to 60% but the Pneumonic type is 100% fatal when left untreated. During Medieval times, this disease caused more than 50 million deaths in Europe, that is about one third of its entire population. Today it affects fewer than 5,000 people a year worldwide but can be deadly if not treated promptly with antibiotics. The disease is transmitted between rodents such as rats and humans via the bite of infected fleas, contact with infected tissues, or inhalation of contaminated respiratory droplets. There is a risk of human plague wherever the presence of the bacteria, an animal reservoir, a vector and human population coexist. This bacterium inspired the strategy of "quarantine" whereby ships, people and goods arriving in Venice in 1485 were held in isolation for 40 days (*quaranta*; Fig. 26). This disease still is with us today, such as in 2017 in Madagascar where it infected 2400 people and killed 200, and in 2024 a resident of Colorado, USA still contracted this bacterial infection.

 | HTTPS://DOI.ORG/10.1515/9783112228265-040

FIGURE 26 The concept of "quarantine" whereby ships were forced to wait offshore for 40 days before entering the city of Dubrovnik, then under Venetian rule, dates back to 1348 when the Great Plague was devastating Europe. Bottom: The bacterium *Yersinia pestis*, causative organism of the Black Plague.

CHAPTER 39

# *Vibrio* and Cholera

## Affecting Three to Five Million People Worldwide Today

Cholera is an acute diarrheal illness caused by infection of the human intestine with *Vibrio cholerae* bacteria. Cholera is caused by a number of bacterial strains, with some types producing more severe disease than others. The genetic variant Serogroup $O_1$ has caused all recent outbreaks, while $O_{139}$ – first identified in Bangladesh in 1992 – caused outbreaks in the past, but recently has been detected only in sporadic cases and never outside Asia [97]. There is no difference in the illness caused by the two serogroups. During the 1854 cholera outbreak in London, John Snow (the father of epidemiology) first suggested that the disease was waterborne and linked to poor sanitation (Fig. 27 [98]). People can get sick when they swallow food or water contaminated with cholera bacteria. When *Vibrio* reaches the gut, the human immune system attacks it, but in response the pathogen releases a toxin that causes the blood plasma to drain into the intestines. This causes explosive diarrhea, projectile vomiting and severe dehydration which if not treated can be fatal for half of the victims. To date, seven cholera pandemics have occurred in the developing world, the first emerged from the Ganges Delta in India in 1817 and subsequently spread by global trade, travel and tourism. In 1831 cholera first arrived in England and in 1892 in Germany where Robert Koch identified the causative bacterium. The most recent pandemic began in 1961 in Indonesia and continues today. Every year the World Health Organization receives reports of some 40 cholera outbreaks. Cholera continues to affect an estimated three to five million people worldwide and causes 30,000 to 130,000 deaths annually. Persons who survive an episode of cholera have long-lasting immunity for at least three years. A number of safe and effective oral vaccines for cholera are available today.

 | HTTPS://DOI.ORG/10.1515/9783112228265-041

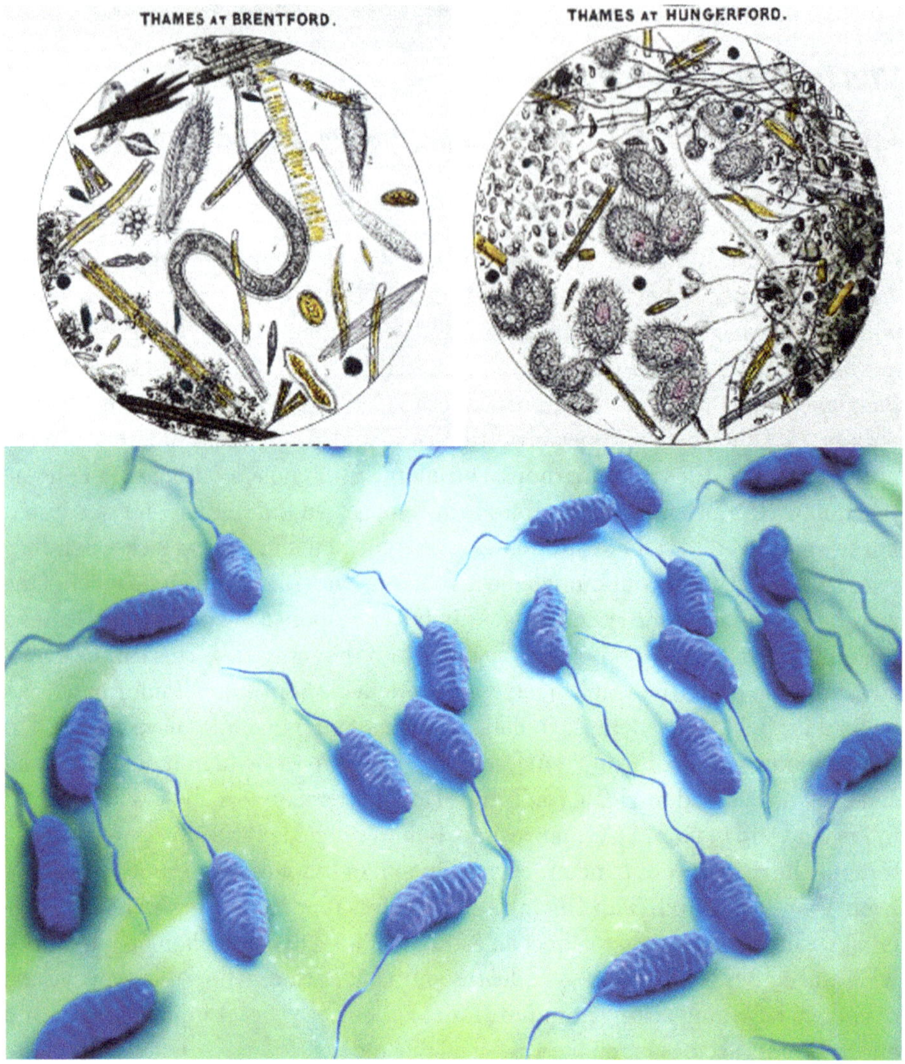


FIGURE 27 Top: illustrations from Arthur Hassall's *Microscopic Examination of the Water Supplied to the Inhabitants of London*, 1850. Bottom: scanning micrograph of *Vibrio cholera* bacteria.

CHAPTER 40

# *Bacillus anthracis* and Biological Warfare

## Source of Feared Biological Weapon

In the wake of the 9-11-2001 terror attack in the United States, 22 people including journalists and members of congress were infected when receiving letters containing white *Anthrax* bacterial powder. Five people died. *Anthrax* bacteria form spores which lie dormant for many years and activate when they come into contact with the human body. *Anthrax* is not contagious from person to person but can be fatal when people inhale airborne spores [99]. In the late 1990s, United States troops fighting in the Gulf War were vaccinated against *Anthrax* because of fear that Saddam Hussein might use biological weapons. In 1972, the United Nations Biological Weapons Convention, signed by 103 countries, made an effort to halt this malpractice.

 | HTTPS://DOI.ORG/10.1515/9783112228265-042

CHAPTER 41

# *Mycobacterium* and Tuberculosis

## World's Top Infectious Killer Today

Tuberculosis is caused by the rod-shaped bacterium *Mycobacterium tuberculosis* which most often affects the lungs – called pulmonary tuberculosis. This disease is thought to have been a serious problem in Egypt as early as 5,000 years ago. The bacterium was discovered in 1882 by Robert Koch, and is spread through the air when infected people cough, sneeze or spit [100]. A person needs to inhale only a few germs to become infected, leading to inflammations (tubercules) which may spread. It is thought that bacteria of the genus *Mycobacterium* initially evolved in soil and that some species adapted to live in mammals. *Myxobacterium tuberculosis* most likely emerged from a cow-adapted species that was much less harmful to humans. *Mycobacterium bovis* causes tuberculosis in cattle and can be transferred to humans via contaminated milk, but this problem has been largely resolved by the pasteurization of milk (heating at 71.5°C for 15 seconds). This process was first developed by French microbiologist Louis Pasteur in the 1860s.

About a quarter of the global human population is estimated to have been infected with tuberculosis bacteria, but only 5–10% will go on to develop the full disease. Those who are infected but not ill cannot transmit it. Famous writers such as Kafka and Camus, musicians such as Chopin and Stravinsky, but also Napoleon and Florence Nightingale all suffered from tuberculosis. People with compromised immune systems, such as those living with HIV, malnutrition or diabetes, or people who use tobacco, have a higher risk. Despite being preventable with a course of antibiotics of streptomycin, every year ten million people fall ill with tuberculosis and 1.5 million people die from this disease each year – making it the world's top infectious killer. The *Bacillus* Calmette-Guérin vaccine derived from tuberculosis bacteria in cattle is used for prevention but does not fully protect against lung tuberculosis. The increasing occurrence of multidrug resistant bacterial strains is adding to the problem. Most of the people who fall ill with tuberculosis live in low- and middle-income countries such as Bangladesh, China, India, Indonesia, Nigeria, Pakistan, Papua New Guinea, Philippines, Timor-Leste, Vietnam and South Africa [101].

 | HTTPS://DOI.ORG/10.1515/9783112228265-043

CHAPTER 42

# *Legionella pneumophila* and Legionnaire's Disease

Legionnaire's disease is a lung disease, first identified at an American Legion convention held in Philadelphia, United States, in 1976, which resulted in 182 illnesses and 29 fatalities. The causative bacterium *Legionella pneumophila* is mostly found in freshwater, cooling towers of air conditioning systems and in soil biofilms. Outbreaks are sporadic but attract significant media attention. Mortality can be <5% if antibiotic treatment is administered swiftly. As a precaution, biofilms in piping system need to be routinely treated with biocides. Similarly, gardeners working with potting soil mixtures need to take precautions of wearing gloves and face masks.

 | HTTPS://DOI.ORG/10.1515/9783112228265-044

CHAPTER 43

# Soil Bacteria

## Important for Agricultural Food Security

A single gram of soil contains billions of bacteria, most of which have yet to be named, and each has its own particular roles and capabilities. Most live in the top 10 cm of soil where organic matter is present. Bacteria play an important role in decomposition of organic materials, especially in the early stages of decomposition when moisture levels are high. *Bacillus subtilis* and *Pseudomonas fluorescens* are examples of decomposer bacteria [102]. Populations of microbes can boom or bust in a few days in response to changes in soil moisture, temperature or carbon substrate. To gain advantage in this process, many microbes release antibiotic substances to suppress competitors. *Bacillus megaterium* and *Pseudomonas fluorescens* are examples of bacteria that have been used to suppress the disease-causing fungus *Rhizoctonia solani*. The bacterium *Bacillus subtilis* has been used to suppress seedling blight of sunflowers, caused by the fungus *Alternaria helianthi*. A number of bacteria have been commercialized worldwide for disease suppression. However, suppression is often specific to particular diseases of particular crops and may only be effective in certain circumstances. *Rhizobium leguminosum* is a soil bacterium that forms a symbiotic relationship with roots of peas and beans (legumes [103]). Beijerinck in 1888 first cultivated this bacterium, then called *Bacillus radicola*, from plant root nodules. These bacteria are able to fix atmospheric nitrogen and convert it to ammonia which is shared with the plant host. They play a critical role in improving crop yield and soil fertility and can add the equivalent of 100 kg of nitrogen per hectare per year. Without them many legume crops would be much reduced and endanger human food security. *Azotobacter*, *Azospirillum*, *Agrobacterium*, *Gluconobacter*, *Flavobacterium* and *Herbaspirillum* are further examples of free-living nitrogen-fixing bacteria. Finally, the characteristic smell of earth after rain is caused by a chemical compound called geosmin produced by selected soil bacteria, cyanobacteria and some fungi. Water itself has no smell, but camels can smell lifesaving water at distances up to 75 km due to geosmin odorant from soil microbes.

 | HTTPS://DOI.ORG/10.1515/9783112228265-045

# PART 6

# Viruses

# Introduction to Part 6

Viruses range in size from 5 to 100 nanometers, that is, hundreds of times smaller than the average bacterium. In 1897, Dutch microbiologist Martinus Beyerinck working on tobacco disease could not detect a bacterial origin, and first postulated involvement of a "*contagious liquid contaminant*" now known as Tobacco Mosaic Virus TMV. Viruses were only visualised in the 1930s when electron microscopes were first used. These smallest of microbes are classified by shape and size but conclusive identification requires molecular sequencing which only started to become widely available from the 2000s [104]. Viruses contain genetic material, either in the form of RNA or DNA, but no known virus contains both. RNA viruses mutate more frequently and hence are more likely to adapt rapidly and jump host species barriers. This explains why so many infectious diseases are caused by RNA viruses such as influenza, measles, polio, Ebola, HIV, AIDS, SARS and COVID. Among the hundreds of thousands of viruses known to exist, most only infect bacteria and only some 260 affect humans. On their own, viruses cannot multiply. For this they need to infect a living host cell, and take over its internal machinery to make thousands of new viruses, often destroying the host cell in the process. Two different life strategies are assumed by viruses. The first one is "hit and run" which is practised by highly infective viruses such as influenza, measles and smallpox which readily transmit to susceptible new hosts. In the second strategy of "hit and stay", the virus achieves some long term symbiotic state in the host, or one in which the host achieves a tolerable disease state. Examples of the latter are Herpes simplex and possibly HIV [105].

Viruses also play a key role in microbial adaption and evolution through the process of horizontal gene transfer. When a virus infects a cell, it inserts its genetic material into that of the host. In this process some viral genes, including those responsible for antibiotic resistance, might be transferred to the host while some of the host genes may be transferred to the virus. This process enables far more rapid adaption than through selective mutations.

Viruses have been found to be associated with every known living organism. Most have a narrow host range, often restricted to a single species. Virus infections often control the abundance of their host, allowing other species to coexist. This strategy is known as 'kill the winner' and is responsible for the maintenance of biodiversity and, for example, the collapse of algal blooms.

 | HTTPS://DOI.ORG/10.1515/9783112228265-046

CHAPTER 44

# *Variola*, Smallpox and Monkeypox

## Killed 800 Million People but Essentially Eradicated Today

Traces of smallpox pustules found on the head of the Egyptian mummy of Pharaoh Ramses V suggest that the disease is at least 3,000 year old. The global spread of smallpox corresponds with the growth of human civilization and exploration. Two forms of the disease are recognized, *Variola minor* with a mortality rate of approximately 1%, and the more common *Variola major* with a higher mortality rate of 30%. Between 65 to 80% of survivors are scarred with pockmarks, most prominently on the face. An early method for controlling smallpox was "variolation". In this treatment, people who never had smallpox were exposed to material from smallpox pustules by scratching into their arm. This led to the development of fever and a rash but much milder than if they acquired smallpox naturally. In 1763 the British army knowingly handed out contaminated blankets among American Indians to decimate their population. In 1796 Edward Jenner noticed that milk maids seldom caught smallpox. He interpreted this to be due to regular contact with the milder symptoms of cowpox acting to confer some form of immunity. This insight ultimately led to the development of vaccines (*vacca* is Latin for cow). While Jenner inoculated people with a closely related cow pathogen, in 1879 this inspired Pasteur to develop the first live vaccine against chicken cholera, based on the observation that chickens previously infected with old bacteria showed no symptoms upon exposure to fresh bacteria. Smallpox only infects humans and is estimated to have killed up to 800 million people during its existence. A global vaccination campaign by the World Health Organization from 1958 to 1977 essentially eradicated smallpox. Many people consider smallpox eradication to be the biggest achievement in international public health. Today, the vaccine is only still being produced as a backup for biological warfare threats [106].

*Monkeypox (MPOX) virus* is related to *Variola* and was first discovered in monkeys in 1958 and the first human case recorded in 1970 in the Democratic Republic of Congo. It can spread from human to human and occasionally from animals such as rodents to humans. A global outbreak occurred in 2022–2023 in 110 countries including the Americas and Europe which affected 87,000 people and caused 112 deaths. In 2024 MPOX was declared a public health emergency of international concern, aided by the fact that collective immunity against poxviruses has deteriorated.

 | HTTPS://DOI.ORG/10.1515/9783112228265-047

*Not to be confused*: Chickenpox is a comparative mild disease (suggested by the name "chicken") but which instead is caused by a member of the herpes viruses *Varicella*. This contagious airborne disease is rarely fatal (one in 60,000 cases), and symptoms are more severe in adults than children. The virus remains dormant in the human body's nerve tissues and in immune-depressed adults can trigger an ailment called "shingles". It can be prevented by vaccination.

CHAPTER 45

# *Morbillivirus* and Measles

## Killed 200 Million People Worldwide but Controlled Today by Vaccination

Measles or Rubeola is caused by the *Morbillivirus* which evolved from rinderpest that infects cattle. Measles can affect anyone but is most common in children. It infects the respiratory tract and then spreads throughout the body. Symptoms include a high fever, cough, runny nose and a diagnostic rash all over the body. The first systematic description of measles, and its distinction from smallpox and chickenpox, is credited to the Persian physician Abu Bakr al-Razi, AD 860–932. At that time, outbreaks were limited and the virus not yet fully adapted to humans. Sometimes between AD 1100 and 1200, the measles virus fully diverged from rinderpest, becoming a distinct virus that infects humans. Measles requires a human population of >500,000 to sustain an epidemic, a situation that first occurred following the growth of Medieval European cities [106].

If measles is continually present in a community, many people develop resistance. However in populations not exposed to measles, exposure can be devastating. In 1529, a measles outbreak in Cuba killed two-thirds of indigenous people who had previously survived smallpox. Similarly measles killed 20% of Hawaii's population in the 1850s and over 40,000 Fijians, approximately 75% of the population. Measles had not been seen on the Faroe Islands in 1846 for 60 years, so almost no residents had any acquired immunity. Three-quarters of the residents got sick, and more than 100 died before the epidemic burned itself out. Globally, between 1855 and 2005, measles is estimated to have killed about 200 million people. Some seven to eight million children are thought to have died from measles each year before a vaccine was introduced. In 1954, the virus was isolated from a 13-year-old boy from the United States and adapted and propagated on chicken embryo tissue culture. Licensed vaccines became available in 1963 and are estimated to have prevented one million deaths per year. Measles as an endemic disease was briefly eliminated from the United States in 2000, but continues to be reintroduced by international travelers and reemerged in 2023. As outbreaks easily occur in under-vaccinated populations, the disease is seen as a test of sufficient vaccination within a human population.

 | HTTPS://DOI.ORG/10.1515/9783112228265-048

CHAPTER 46

# *Influenza* and the Spanish Flu

## Killed 50 Million People Worldwide but Now Effectively Controlled by Vaccination

Influenza is thought to have caused epidemics throughout recorded history. An epidemic is a disease with much higher incidence than expected in a specific location, and we call it a pandemic when it extends to a large geographic area such as a continent or the whole planet. The first probable epidemic occurred in 412 BC, and outbreaks were first called influenza in 1357 (*influenza* is Italian for influence, *flu* for short). Asia likely experienced its first pandemic in 1580 which then spread to Africa, Europe and the Americas. The 1918–1920 Great Influenza epidemic known as Spanish flu, was caused by the highly virulent $H_1N_1$ Influenza A virus. It commenced in the United States but rapidly spread into Europe. The impact of this outbreak was amplified by the cramped living conditions and massive military troop movements during World War I which also reduced peoples' resistance. The virus is thought to have originated from birds, infecting 500 million people (one third of the world's population at that time) and killing 50 million. Throughout Europe, warring countries suppressed bad news on the pandemic to maintain morale, except for newspapers in neutral Spain thus creating the false impression that Spain was the epicenter. In 1933 British researchers Wilson Smith, Christopher Andrewes and Patrick Laidlaw isolated and identified the influenza virus. They found no bacteria in throat washings from patients with influenza but instead discovered that the disease was caused by a virus. Vaccines were attempted but they were mistakenly based on bacteria, not the virus, and thus could only help suppress secondary infections [105]. Researchers have long suspected that different types of influenza viruses existed, as the blood of some influenza patients did not develop antibodies to the strain isolated in 1933. Every flu virus has two types of surface proteins, hemagglutinin (H) and neuraminidase (N), which forms the basis for their taxonomic coding. $H_5N_1$ thus refers to the 5th iteration of H and the 1st iteration of N.

Animal influenza viruses are distinct from human seasonal influenza viruses and do not easily transmit between humans, but some countries have reported cases of human infection from swine influenza. A second pandemic known as the Asian flu emerged in 1957 caused by the $H_2N_2$ *variant*, followed in 1968 by a new $H_3N_2$ *strain* starting in the United States. In 1997 $H_5N_1$ crossed over from birds (*Avian Influenza*), and in 2009 the original $H_1N_1$ strain reemerged but this time from pigs. In 2023 $H_5N_1$ was discovered in the Antarctic for the first time raising concern for impacts on birds such as penguins and skuja not previously exposed to avian influenza viruses. In South America the virus has been responsible for wiping out 40% of the Peruvian

 | HTTPS://DOI.ORG/10.1515/9783112228265-049

pelican population. Scientists also discovered the existence of another strain of the virus: *Influenza B*. In 1942, a new bivalent vaccine was developed that protected against both $H_1N_1$ strains A and B. Each year, different vaccines are now developed, targeting three or four strains of the virus predicted to be most common. In 2026 a new strain termed "Super K", or more precisely influenza A $H_3N_2$ *subclade K*, was causing concern. The endless seasonal cycle of influenza outbreaks in winter and need to adjust strain specific vaccines are estimated to incur an annual economic burden of USD 11 billion in the United States alone. Exposure to influenza early in life can shape long term immunity, thus offering the opportunity for improved vaccination intervention strategies for different birth year cohorts [107].

*Not to be confused*: Influenza damages human lung and bronchial tissues, allowing the normal nose and throat flora to invade the lungs, causing disease and eventually death. Most pneumonia cases at United States Army camps during the 1918 flu epidemic were caused by the bacteria *Streptococcus pneumoniae*, *Haematococcus influenza* and *Streptococcus aureus*. The bacteria *Moraxella catarrhalis* or *Mycoplasma pneumoniae* can also cause pneumonia. Treatment of pneumonia depends upon whether the infection is caused by a bacterium, virus (such as influenza or COVID) or even inhalation of fungal spores of *Aspergillus*, *Cryptococcus* or *Pneumocystis*. The latter fungal category of pneumonia usually only affects people with compromised immune functions.

CHAPTER 47

# *Zika* Virus

The Zika *Flavivirus* is transmitted to humans by the bite of infected *Aedes aegyptii* mosquitoes that thrive in stagnant waters. This virus was first identified in monkeys from the Ziika forest of Uganda in 1947, and five years later observed in humans in Uganda and Tanzania. Most people do not develop symptoms, but those who do, develop rash, fever, conjunctivitis, muscle and joint pain, and headache lasting for two to seven days. Rarely, autoimmune disorders affecting the nervous system have been reported. However, of considerable concern is that the geographic regions affected experienced a rise in the numbers of babies born with a smaller than normal head (microcephaly). The first reported Zika outbreak occurred in 2007 on the Pacific Island of Yap (5,000 people), spreading to other Pacific Islands in 2013 (30,000 people), until in 2015 Brazil was hit the hardest (1.6 million cases). Rapid spreading to 89 countries almost certainly was mediated by travelers, with only a small proportion of illnesses contracted from local mosquitoes. No vaccine is available to date. The only protection is mosquito control, and for pregnant women avoiding travelling to danger areas in the Americas. In total some 900,000 cases were reported, with an estimated 51 fatalities. By late 2016 the World Health Organization declared the end of this pandemic.

In 2024 several cases of the related Oropouche Virus (OROV) responsible for "sloth fever" were reported from European travelers who returned from Cuba and Brazil. This disease which is endemic in the Amazon Rainforest is mainly spread by mosquitoes and originates in sloth bears. Similar to the Zika Virus, pregnant women were cautioned due to the risk of miscarriage and birth defects.

 | HTTPS://DOI.ORG/10.1515/9783112228265-050

CHAPTER 48

# *Pilchard Herpes Virus* (PHV)

## Example of Damage from Exotic Viruses to Marine Fisheries

In 1995 in South Australia, a mass mortality of pilchards *Sardinops sagax* which started near the Eyre Peninsula, then progressed in coastal waters both to the east up to Sydney and west up to Perth [108], was ultimately associated with *Pilchard Herpes Virus PHV* [109]. The focal origin, fast rate of spread, high mortality rate (60% of fish stock lost) and lack of previous mortality events of comparable magnitude led scientists to suggest that the disease agent was most likely an exotic pathogen to which Australian pilchards were naïve. Introduction of a foreign pathogen with frozen Californian pilchards used to feed aquacultured tuna in Port Lincoln, South Australia, was suspected but never conclusively proven. A second mass mortality wave originated in South Australia in early October 1998 and similarly spread across the full pilchard species range through southern Australia, over a period of seven months. Pilchard stocks ultimately recovered to normal after a period of several years.

 | HTTPS://DOI.ORG/10.1515/9783112228265-051

CHAPTER 49

# *Ostreid Herpes Virus* and Pacific Oyster Mortality

## Example of Cryptic Aquaculture Disease Triggered by Unusual Environmental Conditions

Pacific Oyster Mortality Syndrome (POMS) is a viral disease which only affects Pacific oysters (*Crassostrea gigas*). It is caused by the *Ostreid Herpes Virus* (*OsHV-1*) and triggers rapid death and high mortality rates – up to 100% within days – in farmed Pacific oysters. There are no human health concerns from consumption of infected oysters. Pacific oysters were intentionally translocated around the world in the 1940s from Japan into Europe, Australia and New Zealand. In 2008 the Pacific Oyster industry in France was the first to be affected by POMS, followed in 2010 by New Zealand. In Australia POMS was first detected in the Sydney region in 2010 and later in Tasmania in late January 2016, in association with a marine heatwave event (>20°C). In addition to shellfish stock losses, the major financial impact was caused by a ban on traditional oyster spat movements from Tasmania to farmers in South Australia and New South Wales with long lasting percussions for the entire AUD 53M Australian oyster industry.

Both above examples illustrate how single species aquaculture industries that involve geographic translocation of brood stock or even bait fish are highly vulnerable to exotic disease outbreaks. Fish maintained in monoculture are more vulnerable to infections for reasons such as higher stocking densities, infected fish are not removed immediately, inadequate nutrition, poor water quality, and handling practices stressful to fish. Fish farm diseases can result from bacterial infections, viral diseases but also fungal pathogens. The relationship between the pathogen, host, and environment determines the disease onset and progression. Stressful conditions, such as sudden temperature changes and hypoxia, can accelerate the spread of pathogenic bacteria and cause severe disease outbreaks. The excessive use of antibiotics such as tetracycline, oxytetracycline and chloramphenicol in finfish aquaculture, which can also contaminate wild fish, poses similar threats as those for human health in inadvertently selecting for resistant superbugs [110].

 | HTTPS://DOI.ORG/10.1515/9783112228265-052

CHAPTER 50

# *Apthovirus* and Foot and Mouth Disease in Cattle

## Major Economic Threat to the Global Livestock Industry

Foot and mouth disease is a highly contagious viral disease that affects all cloven-hoofed animals such as buffalo, camels, cattle, deer, elephants, giraffes, goats, lamas, pigs and sheep. Horses are not affected. The disease can be carried in frozen, chilled and freeze-dried meat and dairy products, as well as in soil, bones, untreated hides, vehicles and equipment used with these animals. The causative virus belongs to the genus *Apthovirus* and includes seven serotypes: O, A, C, $SAT_1$, $SAT_2$, $SAT_3$, and Asia. Infection with any one serotype does not confer immunity against other serotypes. Symptoms include blisters on the mouth, snout, tongue, lips or between and above the hooves on the feet. Most diseased animals will recover in about two weeks. Mortality is normally <5%, but may be higher in young animals. Humans do not get this animal disease, but our concerns relate primarily to reduced meat and milk production. This problem triggered the creation of the first Veterinary School in France in 1762. Vaccines treat the symptoms but animals still carry the virus. This leaves strict quarantine protocols as the key solution to curb its spread. Outbreaks are common in parts of Africa, the Middle East, Asia and South America. Recent outbreaks in the United Kingdom in 2001 and 2007 resulted in millions of animals being culled and billions of dollars of revenue lost due to restriction on meat exports from affected areas. In May 2022, an outbreak was reported in cattle in Indonesia which put neighbouring Australia on alert.

*Not to be confused with*: Hand, foot, and mouth disease in children is often confused with foot-and-mouth disease. Humans do not get the animal disease, and animals do not get the human disease.

 | HTTPS://DOI.ORG/10.1515/9783112228265-053

CHAPTER 51

# *Leporipoxvirus* and Myxomatosis

## Critical Biological Control for Australian Fight against Introduced European Rabbits

Myxomatosis is a disease affecting rabbits but people, dogs, cats, birds or humans are not at risk. It is caused by the Myxoma virus, which is a pox virus of the genus *Leporipoxvirus*. While a mild disease in its native South America for tapeti (a species of cottontail rabbit), it usually is fatal for European rabbits. The virus is spread by blood sucking mosquitoes and rabbit fleas, but once infected the rabbit can transmit to others via secretions from the eyes and nose.

When European settlers naively introduced rabbits into Australia in 1859 for hunting pleasure, this led to disastrous consequences for sheep farming, agriculture and soil ecology. To combat this problem in 1950 Australian federal government scientists deliberately introduced myxomatosis. By 1952–1953, Australia's wool and meat production recovered up to AUD 68 million. However, what initially looked like a success story, with 99.8% of rabbits affected in the first year, and a 90% death rate in the second year, this was reduced to only 25% rabbit mortality seven years later. Rapid genetic selection had occurred among rabbits to generate resistant viral strains.

In response, in 1995 a second virus, Rabbit Hemorrhagic Disease Virus also known as *Rabbit Calicivirus*, was introduced in an Australian field trial. During these trials the virus escaped putatively via bushflies and/or mosquitoes. In the end the rabbit calicivirus did its job as intended, so far without any adverse impacts and major improvements to agricultural production and soil ecology. Scientists are well aware that in future rabbits could also become resistant to calicivirus. The Australian rabbit war therefore is likely to continue indefinitely [111].

 | HTTPS://DOI.ORG/10.1515/9783112228265-054

CHAPTER 52

# Bacteriophages

Phages are "good" viruses that infect bacteria but are harmless to humans, plants and animals. In medicine, they can be used to tackle a single bacteria species, or multiple species simultaneously; or they can be used in concert with antibiotics for optimized medical treatment [112]. As an alternative to antibiotics such as penicillin, phage technologies are currently actively being explored to harness nature to engage in inter-species battle and pit microbes against microbes. Advantages of phage therapies are their specificity, they are cheap to produce and phages disappear when the host is killed. Problems to be solved include that bacteria can also develop resistance against phages, although many phages rapidly evolve to overcome this resistance, and that phages may be difficult to deliver to internal organs to be treated. Let's hope we get it right this time! [113].

 | HTTPS://DOI.ORG/10.1515/9783112228265-055

CHAPTER 53

# *Human Immunodeficiency Virus* and AIDS

## Responsible for One of the Deadliest Post-World War II Epidemics

Acquired Immune Deficiency Syndrome (AIDS) was first recognized as a new disease in 1981 when increasing numbers of young homosexual men succumbed to a rare form of pneumonia and rare vascular cancer (Kaposi sarcoma). Ultimately, in 1986 this disease was linked to two lentiviruses, human immunodeficiency viruses 1 and 2 (*HIV-1* and *HIV-2*) which were demonstrated to have been caused by multiple cross-species transmissions of simian immunodeficiency viruses (SIVs) from African apes [114]. AIDS is caused by an RNA virus which uses as its host white blood cells (T4 lymphocytes). Surprisingly, these viruses appeared to be largely non-pathogenic in their natural primate hosts. At the end of the 19th century, the German physician Rudolph Virchow coined the term *zoonosis* (*zoo* = animal; *nosos* = disease) for infectious diseases transmitted from animals to humans (Fig. 28). The HIV virus evolves around one million time faster than mammalian DNA because it has such short generation time and the reverse transcription process, they use for copying their genetic material is error prone. The AIDS pandemic has infected at least 60 million people and caused more than 36 million deaths, making it one of the deadliest epidemics in post-World War II history. Antiretroviral (ARV) drugs have reduced the toll of AIDS-related deaths, but access to therapy is not universal (costing USD10,000 per year per patient) with the highest prevalence in young adults in sub-Saharan Africa. AIDS is primarily a sexually transmitted disease and heralded the end of the "free love" attitude of the 1960s hippie culture and put a chill on intravenous drug taking. Luckily, any HIV virus which a mosquito sucks up would be killed by the mosquito's metabolism. When the day comes that the virus mutates to survive this, we could be in real trouble [115].

 | HTTPS://DOI.ORG/10.1515/9783112228265-056

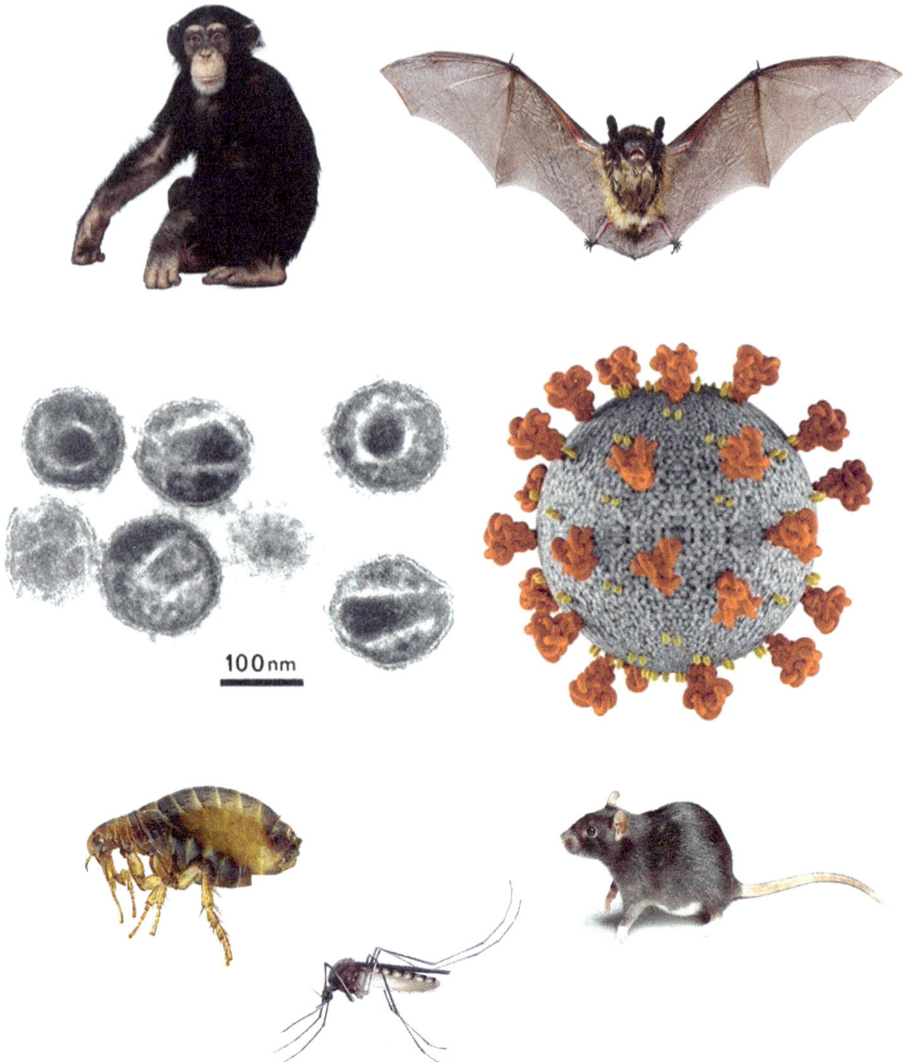


FIGURE 28 When humans encroach on animal habitats that harbor viruses and bacteria, this can cause microbes to jump species barriers and infect humans, e. g., AIDS virus originated from chimpanzees and COVID virus most likely from bats. Fleas and rats can play a role in transfer of microbes to humans, notably with *Yersinia pestis* bacteria responsible for the Plague, and Malaria *Plasmodium* protists transferred by mosquitoes.

CHAPTER 54

# Coronaviruses and COVID-19

## The 2019 Pandemic Killed Seven Million People and Had Far Reaching Mental Health, Political and Socioeconomic Consequences

Coronaviruses are a large family of viruses named for the club-shaped spikes projecting from their surface as in a crown (*corona*). Some species cause the common cold in people, others infect animals such as bats, camels and cattle. Scientists first identified a human coronavirus in 1965 which caused the common cold. The virus that causes SARS (*Severe Acute Respiratory Syndrome*) emerged in southern China in 2002 and quickly spread to 28 other countries. It is thought to have spilled over via palm civets from cave-dwelling horseshoe bats. SARS coronavirus causes fever, headache, and respiratory problems such as cough and shortness of breath. More than 8,000 people were infected by July 2003, and 774 died. MERS (*Middle East Respiratory Syndrome*) started in Saudi Arabia in 2012 spilling over from dromedary camels. It causes the same respiratory symptoms but can also cause kidney failure. Almost all of the nearly 2,500 cases occurred in people who lived in or traveled to the Middle East. The MERS coronavirus was less contagious than its SARS cousin but more deadly, killing 858 people.

Humanity will not forget for decades to come, when in December 2019, a pneumonia-like outbreak was first reported in Wuhan, China, traced to a novel strain of coronavirus, which was given the name SARS-CoV-2 or COVID-19 for short. The virus has a 96% similarity to a bat coronavirus RaTG13 isolated from a bat cave thousands of kilometers away from Wuhan. While there is no evidence that this virus was genetically modified, neither bats nor pangolins were sold at the Wuhan wet market, and its precise origin (nature or laboratory leak) remains to be fully resolved [116,117]. Of note is that the early SARS virus had a higher fatality rate (12%) than COVID (2%) but was less contagious. So far, new COVID viral strains such as *Omicron, Pirola, BA2.86, JN.1, FLiRT, FLuQE* have all proven to be less harmful than the original 2019 strain. As of September 2023, there had been 770 million confirmed human infections and seven million confirmed deaths in what set off a global pandemic with devastating impacts in terms of mental health, political unrest and socioeconomic hardships. However, to put those numbers in a global perspective, less than 1% of the seven billion human population in 2019 suffered from illness and the fatality rate was 1–2%. On the scale of pandemics, COVID was moderate (Table 1). COVID taught us the benefits of wearing face masks (long practised in China and Japan), and to work from home via Zoom instead of commuting every day to office. Medical science was lauded for developing vaccines in record

 | HTTPS://DOI.ORG/10.1515/9783112228265-057

speed (13 billion doses were administered, including newly designed mRNA vaccines), but at the same time it created Rapid Antigen Test (RAT) billionaires, and vaccine apartheid between rich and poor countries [118]. We also cannot ignore the unprecedented global plastic waste generated from the need for single-use face masks and RAT tests (Fig. 29 [119]). Plastics were initially designed to protect us against microbes but now we desperately need their help to break down our waste.

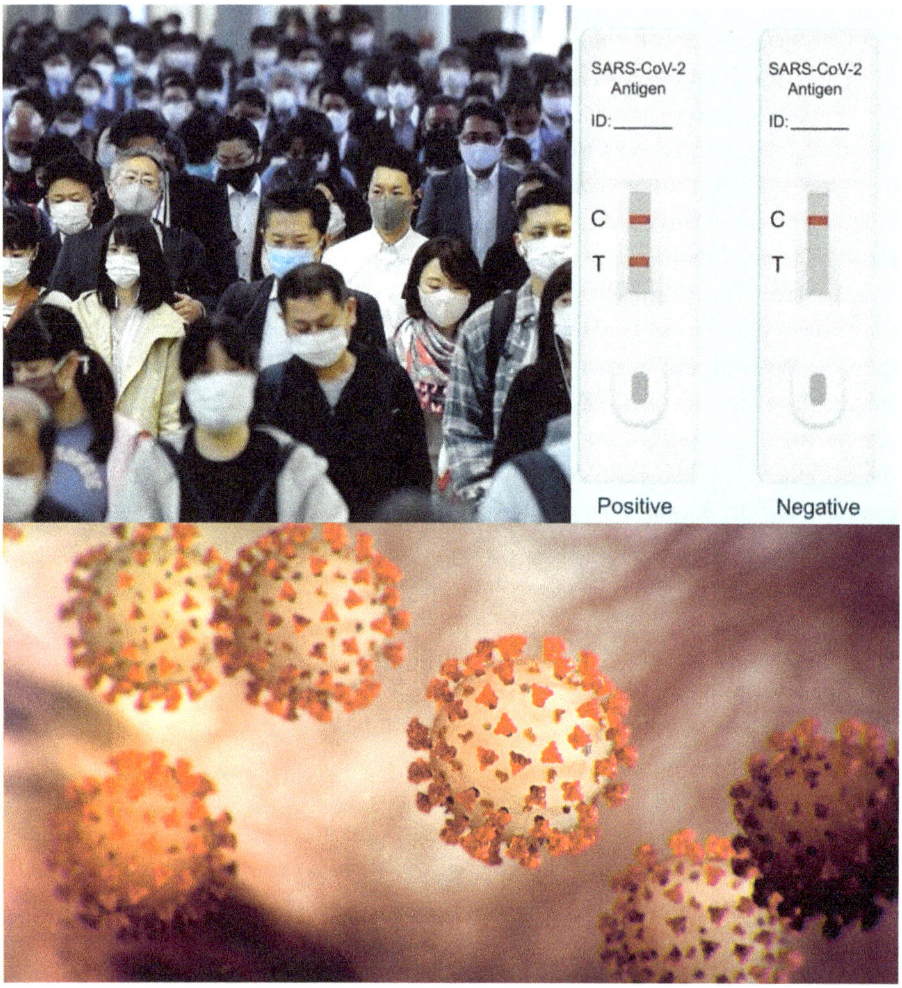


FIGURE 29 The COVID-19 pandemic made us familiar with face masks, rapid antigen tests, and the coronavirus.

PART 7

# What Can We Learn from Microbes?

CHAPTER 55

# The Microbial World Darwin Never Knew

Charles Darwin's famous insights in 1859 into *"The Origin of Species"* focused on harsh competition between large plants and animals on the land, where lions and antelopes struggle to coexist in the same locality (*Survival of the Fittest*). Classical laws of competition claim that when two species compete for the same resource, ultimately only one will persist and the other driven to extinction. However, Darwin never knew about the microbial world on our planet [120]. In 1961, the freshwater planktologist Evelyn Hutchinson [121] pondering the diversity of microscopic life in a small pond near the hilltop church of Santa Rosalia in Sicily, posed the famous question known as the *Paradox of the Plankton*: *"How is it possible for a number of species to coexist in a relatively unstructured environment all competing for the same sorts of materials"*. The answer lies in the fact, firstly, that the water environment is highly structured, both in time and space, composed of continuously changing pulses and patches of light, nutrients, temperature, and animal grazing. The environment is never in equilibrium and no single species therefore can dominate for long. Other scientists pointed out that at microscopic scales these organisms are in reality living kilometers apart [122]. Secondly, different species of microbes have vastly differing resource requirements, allowing them to coexist. Thirdly, fast growing microbes are able to adapt very quickly to the ever changing environment. This concept of *Paradox of the Plankton* equally could have been called *Paradox of the Microbes*. Environmental disturbances caused by nutrient pollution, host abundance, or invading foreign strains can briefly cause an upset, such as an algal bloom or microbial disease outbreak, but the ecosystem rapidly responds to create a new equilibrium. Most algal blooms are a temporary response to a nutrient pulse or escape from grazer control, or the impact of viruses to "kill the winner" [123]. We humans naively fear that a microbial pandemic such as COVID acts as a competitor that aims to kill us all. Instead, most disease outbreaks will peter out to create a new world order until a new genetic strain emerges, and the game of microbial roulette starts all over again (Fig. 30).

 | HTTPS://DOI.ORG/10.1515/9783112228265-058

FIGURE 30 Most human disease pandemics start with incomplete knowledge and fear leading to misinformation and scare campaigns. The 1987 Grim Reaper TV commercials in Australia demonized the broader LGBTQ+ community, but today with major scientific advances in testing, prevention and treatment, living with HIV is no longer a death sentence and our outlook on this problem has changed remarkably.

CHAPTER 56

# Microevolution: The Power of Growing Fast

The average generation time for humans during the past 250,000 years, that is the age when we produce offspring, has been 26.7 years [124]. That means that any genetic adaptations resulting from sexual recombination of the genes of our two parents or the outcome of very rare successful mutations are occurring at a snail pace, and hence are no match for new tricks by fast growing and fast evolving microbes. One well-studied example is the relation between humans and malaria. Malaria parasites already existed in Africa around 100,000 years ago long before humans evolved, but a highly virulent parasite strain emerged some 10,000 years coinciding with the beginnings of agricultural settlement and increased human population. Molecular evidence has demonstrated how this exerted a powerful selective force for the genetic divergence of African, Asian, South American and European human populations. To date only the original sub-Saharan African humans are completely resistant to malaria [125].

Typical microalgae can divide up to once per day up to every three days, that means starting with one cell and assuming no losses from death or being eaten, one ends up with an algal bloom of 300 million cells in one month. The reasons that our oceans are not becoming greener is simply because microalgae are continuously grazed away. Bacteria can reproduce every 20 to 30 minutes that is we may have ten billion cells at the end of a single day. In the same period a human body cell can just manage to produce a single division. Bacteria carry only one set of genes (not a double set as in humans) and simply divide, but never fuse with each other like eggs and sperm do. Even so mutations are rare – on the order of one error per 1,000 rounds of replications, this creates an ample number of variants to adapt to new environmental conditions. On top of that, bacteria have other tricks on their sleeves, they can take up DNA from the environment as if it were "food" or transfer genes horizontally on to companion bacteria, as distinct from vertically through ancestral lineages [120]. By analogy, it is as if a human could go to an insect to steal the genetic coding to sprout wings [115].

Viruses also contain genetic material, either in the form of RNA or DNA, but no known virus contains both. Viruses replicate in eight to 72 hours. RNA viruses mutate more frequently and therefore are more likely to adapt and jump species barriers. This explains why so many diseases are caused by RNA viruses such as influenza, measles, polio, Ebola, HIV, AIDS, SARS and COVID. The HIV virus evolves around one million times faster than human DNA because it has such short generation time and the reverse transcription process they use for copying their genetic material is error prone [114]. In conclusion, we humans are way too slow to beat microbes and we need to work faster and smarter to understand them.

 | HTTPS://DOI.ORG/10.1515/9783112228265-059

CHAPTER 57

# Microbes and their Toxins

Toxins are produced by bacteria, fungi, microalgae, protozoa and, least studied, viruses. The reasons why these organisms produce toxins remain poorly understood, but one thing is clear. Microbes and their chemical arsenal evolved long before humans turned up. They never meant to harm us. These chemicals play a wide range of roles for the organisms themselves, such as signaling between friends, chemical warfare between competitors, nutrient uptake or storage, killing prey or deterrent against being eaten, and others. Some of the most potent toxins are produced by the more primitive bacteria such as botulinum toxin (by *Clostridium botulinum*), tetanus toxin (by *Clostridium tetani*), and tetrodotoxin (thought to be produced by *Vibrio*). All three bacterial toxins block sodium channels in mammalian nervous systems and thereby cause muscular paralysis that can be fatal for humans. Cyanobacterial toxins can cause liver damage (heptatotoxins such as microcystins), skin irritation (dermatitis) or neurological problems (anatoxins, saxitoxins, cylindrospermopsins). The microalgal toxin domoic acid plays a role in iron nutrient uptake by microalgal diatoms and can be stimulated by animal plankton grazing pressure. However, an unintended impact is the way it interferes with brain function in seabirds (Hitchcock's *the Birds* movie inspiration) and in mammals such as humans, dolphins, whales, sealions and manatees. Dinoflagellates which occupy a primitive position among the protists are masters of producing a wide range of neurotoxins impacting on human sodium channels (saxitoxins, brevetoxins) and calcium channels (ciguatoxins) and protein phosphatases (okadaic acid causing diarrhetic shellfish poisoning). Fish killing algal toxins mostly cause pores in the sensitive fish gills (e. g., karlotoxin, prymnesins) but often have no known human health impacts. The natural function of most aquatic toxins is unclear because, with our usual narcissism, research has primarily focused on the impacts these compounds have on humans. Fungal toxins are highly diverse and can contaminate a wide variety of agricultural crops. Aflatoxin- and fumonisin mostly affect humans with weakened immune systems or people that eat food contaminated with moulds. Microbial toxins can be foes, but also have important uses in medical science and research. Botulinum toxin, first linked to sausage poisoning in Germany in the 1900s, nowadays is associated with cosmetic anti-wrinkling applications with global sales of USD 3B per annum for Botox. The role of penicillin and streptomycin as antibiotics in medicine similarly represents a USD 25B per annum industry. Defense interests in bacterial toxins such as anthrax or dinoflagellate toxins such as saxitoxin (and many others studied in secret military laboratories) are also noted. For many potentially toxigenic microbes, we know they can occur as toxic and non-toxic strains, but we still understand very little of what triggers the

 | HTTPS://DOI.ORG/10.1515/9783112228265-060

"virulence factor". The local emergence of new highly virulent genetic strains causes the greatest problems for human society. Dutch microbiologist Baas-Becking in 1934 famously claimed that "*everything is everywhere, but the environment selects*" [126]. We no longer believe this is correct, especially for microbes. The classical species concept as we know it from land plants and animals (dating back to Carl Linnaeus in 1735) does not apply to microbes. What looks the same, may well hide very different organisms. For example, the famous bacterium *Escherichia coli* can be found in the large intestines of most healthy human beings, but only selected strains (serotypes) can cause bacterial diseases such as diarrhea, sepsis and urinary tract infections. While the widespread coccoid, minute ocean cyanobacterium *Prochlorococcus* may not show any morphological diversity, almost every strain that has been genetically sequenced appears unique. Similarly, morphologically indistinguishable *Alexandrium* dinoflagellate strains include non-toxic and toxic clades. We are well familiar with this concept in the context of influenza and coronavirus pandemics. Most pandemics or blooms are multi-species or multi-strain events. Similarly, quarantine as a management strategy against harmful microbes needs to be targeted at the genetic strain level.

CHAPTER 58

# Quarantine for Ships, Aquaculture and Agriculture

Less well known is that in addition to human health impacts, unwanted spread of harmful microbes can impact fisheries, aquaculture and agriculture. Ship ballast water [127] has been claimed to have introduced toxic dinoflagellate blooms into areas where they did not exist previously, but also been demonstrated to be a carrier in the spread of cholera bacteria. Ballast water was first suggested as a potential vector for transport of unwanted marine microorganism over 100 years ago. However, it was only in 2004 when the International Maritime Organization (IMO) adopted the International Convention for the Control and Management of Ship's Ballast Water and Sediments, but which took until 2017 to become formally ratified [128] . The Convention forces all merchant ships to install ballast water treatment systems on board (including heating and biocides) to manage their ballast water and sediments to a prescribed microbial standard. For organisms smaller than 10 µm, discharged ballast water should not contain more than *Vibrio cholerae* (serotypes $O_1$ and $O_{139}$) one colony-forming unit per 100 ml; or *Escherichia coli* <250 colonies per 100 ml; and for *Enterococcus* species <100 colonies per 100 ml. Considering that global shipping moves 10 billion tons of ballast water around the world annually, it is surprising that this has not caused more problems for ocean microbes or simply we are not yet aware of the ecosystem damage hidden beneath the waves. Increasingly strict guidelines are also in place to prevent the spread or inadvertent introduction of aquaculture diseases. Visitors of aquaculture or agriculture operations are well-familiar with disinfectant boot cleaning stations, and this similarly applies to people and cars to keep out *Phytophthora* "root rot" from Australian wilderness areas. Outbreaks of foot-and-mouth disease in cattle can trigger major trading barriers and economic hardship. Just like the consequences of COVID barriers for global trade, abuse of quarantine barriers can and has been applied unfairly for trade advantages. Since 2010 increasingly strict quarantine as well as intellectual property regulations also apply to the exchange of scientific samples and microbial cultures between collaborating scientists, to ensure the country of origin attracts a fair share of commercial benefits from the use of its genetic resources (Nagoya protocol [129]). The manner in which in the 1940s Howard Florey rubbed *Penicillium* fungal spores into his clothing, so that, if during the war the laboratories were destroyed or any of the scientists killed, the research could start elsewhere again, would no longer be possible today. Bureaucratic solutions need to found for the necessity for rapid exchange of research samples and molecular sequence data in the event of an imminent global emergency such as COVID-19. Unresolved issues exist around gene ownership, including those discovered from ocean microbes, with as of today ten nations owning close to all marine gene patents.

 | HTTPS://DOI.ORG/10.1515/9783112228265-061

TABLE 1 Summary of the top nine microbial diseases that have impacted on human society, arranged from highest to lowest fatalities. Both the Plague and Spanish Flu in Medieval times occurred when no vaccines were available and killed an estimated 30% of the population of Europe. By comparison, while COVID-19 caused 770M illnesses, this constituted only 1% of the 7.8B human population in 2019, and fatalities amounting to 7M were <1% of total illnesses thanks due to massive vaccination and quarantine efforts.

| Disease | Causative organism | Zoonotic origin | First discovered | Illnesses/ control | Human fatalities |
|---|---|---|---|---|---|
| **Plague (Black Death)** | bacterium *Yersinia pestis* | Spread by fleas and rats | 1894 Alexandre Yersin | Today <5,000/yr (Quarantine) | 50M (30% population Medieval Europe) |
| **Influenza (1918 Spanish Flu)** | $H_1N_1$ virus (RNA virus) | birds or swine | 1933 Smith, Andrewes & Laidlaw | 500M (Vaccination) | 50M (30% population Medieval Europe) |
| **Smallpox** | *Variola* virus (DNA virus) | related to cowpox | 1796 Edward Jenner | Today nearly eradicated (Vaccination) | 800M (total) |
| **Measles** | Morbillivirus (RNA virus) | rinderpest | 860–932 Al-Razi | Today vaccination control | 200M (total) |
| **AIDS** | HIV-1 and HIV-2 (RNA virus) | African primates | 1981 Gallo, Montagnier & Levy | 60M; antiretroviral drugs | 36M |
| **COVID-19** | SARS-CoV-2 virus (RNA virus) | ? bat coronavirus RaTG13 | 2019 Zhang Jixian | 770M (<1% global population) (Vaccination/ Quarantine) | 7M (<1% illnesses) |
| **Malaria** | protist *Plasmodium falciparum* | *Plasmodium* in gorillas. Mosquito borne | 1902 Ronald Ross | 300–500M/yr; (Quinine, Artemisinin) | 2.7M/yr |
| **Tuberculosis** | bacterium *Myxobacterium tuberculosis* | ? soil or cattle | 1882 Robert Koch | 10 M/yr (5–10% lead to illness) | 1.5M/yr |
| **Cholera** | bacterium *Vibrio cholerae* | waterborne | 1854 John Snow | 3–5M/yr (Sanitation) | 30,000–130,000 /yr |

CHAPTER 59

# Quarantine for Human Disease Outbreaks

Human disease outbreaks invariably result from the introduction of a new pathogen into naïve host population never previously exposed to such threat (Table 1). When Christopher Columbus explored the "New World" he inadvertently did not only bring sailors with him but also invisible microbial stowaways that caused devastating European diseases such as smallpox, measles and typhus into the Americas. It was not Spanish soldiers that conquered the Incas and Aztecs but microbes [130]. Three quarters of emerging human diseases have confirmed origins in animals, either wildlife such as bats, primates, rodents and birds, or domesticated animals such as poultry, pigs and camels. Measles originated from rinderpest, smallpox is related to cowpox, AIDS originated from primates, COVID from bats. Changes in land use, deforestation, agricultural expansion, trade in wildlife and ever expanding human settlements are escalating the impacts on humans. Climate change expands the range of diseases transmitted by mosquitoes such as malaria and yellow fever. Increased ease of global plane travel makes spread of airborne disease more likely than ever. Whereas in the 1330s the Black Death took eight years to move from Kyrgyzstan to Crimea, and in 1918 it took Influenza three months to move from Kansas to Europe, in 2019 COVID spread from Wuhan to Europe and North America in a time span of a mere three weeks. The term quarantine was first coined in the 14th century when ships arriving in Venice from plague infected ports were required to sit at anchor for *quaranta* (40) days before permission was given to land. It is of interest to compare ad hoc counter measures against the Spanish flu in the 1920s with those put in place in 2019 against COVID. Border closures for island nations such as Iceland, Australia, and American Samoa saved many lives. Social distancing measures were introduced such as banning mass gatherings and closing schools, theatres, and churches, and limiting public transportation. Wearing face masks became common already then in some places such as Japan, but even then, there were debates over their efficacy and resistance to their use. Early fabric face masks used in 1918 were not fine enough to trap something as tiny as a virus. Spread of misinformation and bans on reporting human death tolls were also rife, well before social media escalated this problem to problematic proportions in the 2020s. Vaccines were attempted in 1920 but they were mistakenly based on bacteria and not the virus, and thus could only help with treating secondary infections.

The use of quarantine and other measures for controlling epidemic human diseases has always been controversial because such strategies raise political, ethical, and socioeconomic issues and require a careful balance between public interest and individual rights [131]. Quarantine practices are now the cornerstone of organized responses to infectious disease outbreaks. However, nobody would like to see

 | HTTPS://DOI.ORG/10.1515/9783112228265-062

us return to politicians showing off their strong-handed leadership skills by imposing harsh COVID-zero measures. A zero quarantine risk simply is not achievable, but not doing anything is also not acceptable. Retrospective evidence has demonstrated that recovering from natural COVID infection produced better immunity then vaccination [132]. This suggests that there may be important advantages to be gained if future vaccination and therapy are fine-tuned toward individual patient's immunological profiles to achieve faster and stronger herd immunity (as advocated by countries such as Sweden). Elderly people or those with depressed immune systems would be well advised not to travel around the globe without reintroduction of vaccination passports to be inspected prior to boarding flights, or better just stay home. On the other hand, people with certified strong immune systems could be given the opportunity to apply for exemptions from blanket enforcement of quarantine lockdowns, not only for their own mental health but for the benefits of continuation of human society as a whole. When the Human Genome Project was completed in 2003, we were all promised that "personalized medicine was just around the corner". In the same way that food, water, and energy security for an ever increasing human population have forced us to take a step back from moves toward globalization, microbial biosecurity has raised its head as an additional border patrol hurdle to be reckoned with. Existing national government structures no longer serve planetary interests, and we may need to radically rethink a structure whereby the World Health Organization takes charge of pandemics and the Intergovernmental Panel on Climate Change becomes responsible for climate mitigation [133].

CHAPTER 60

# Microbes and Climate Change

Microbes are major players on our planet whose activities can exacerbate global change but may also provide powerful novel solutions to problems arising due to anthropogenic activities. While the impact of human activities and climate change on loss of plant and animal biodiversity has been well publicized, their impact on microorganisms has not been widely discussed. Ease of global travel and human population growth increase the spread of animal, human and crop pathogens, and better understanding microbial community ecology is key to developing strategies for pathogen control. Climate change and population growth increase antimicrobial resistance and at the same time exacerbate the global spread of vector-borne pathogens and their diseases. Anthropogenic climate change increases diseases caused by crop pathogens, but also stresses marine life, causing disease and disrupting normal ecosystem function, thereby threating global food security of both agricultural and seafood crops. Microbes affect every aspect of our human lives and we need greater microbial literacy within human society and our leaders to ensure we make the best decisions for our planet [134]. Examples are whether to give birth by Caesarean or natural delivery, breast-feed infants, be vaccinated or treated for an infection, use germicidal soaps and household cleaning products, acquire a companion household pet [135], where to holiday, or what food to eat and how to store and prepare it.

 | HTTPS://DOI.ORG/10.1515/9783112228265-063

# Epilogue

I hope to have convinced the reader that our planet belongs to microbes. Humans are just mere passengers in a train travelling through the amazing land and seascapes that microbes have created. It is not just the organisms but also the processes they created. We should be in awe not only of the singers but also their songs. Treating microbes as our enemy with agricultural biocides, household disinfectants, overprescription of antibiotics and vaccines does no longer serve us on this overcrowded planet. Microbes are not out to kill humans, and we need to work smarter and faster to understand who they are and what they are doing here. Classical ecological concepts which originated from studying large land animals and plants do not apply to the microbial world of very fast growing creatures that continuously adapt to a changing environment. Any disturbances that humans may create such as nutrient pollution triggering fish-killing algal blooms, climate change expanding the geographical range of malaria mosquitoes, or deforestation driving wildlife diseases into human settlements, ultimately lead to a new equilibrium. Most of us feared for COVID to kill us all, but no, by the time this virulent virus had distributed itself over the planet its job was done and it petered out. Letting this microbe escape out of China on to an unsuspecting Western world never previously exposed to this type of SARS coronaviruses has been largely inflicted by our appetite for global travel. Quarantine border closures in combination with improved sanitation and wearing of face masks are effective counter measures. Elderly people or those with depressed immune systems would be well advised not to travel around the globe without reintroduction of vaccination passports to be inspected prior to boarding flights (as still is in place for e. g., yellow fever). Advances in our understanding of the human gut microbiome newly revealed how our "hygiene revolution" of excessive household sanitation has reduced human exposure to "old friends", that is ancient microbes in mud and rotting vegetation that co-evolved with mammalian immune systems in isolated hunter-gatherer and farming communities. We no longer need to fear microbes because we cannot see them. Increased access to molecular techniques can readily identify who they are, their geographic or zoonotic origin, elucidate vectors of spreading, their interactions with our immune system and how best to manage their impact. We need to live in symbiosis with them, rather than be at war. Let's learn from them while we still can!

 | HTTPS://DOI.ORG/10.1515/9783112228265-064

# Glossary

**Acquired Immune Deficiency Syndrome** potentially life-threatening condition caused by the human immunodeficiency virus (HIV) which attacks the immune system. Antiretroviral drugs have reduced the toll of AIDS-related deaths, but access to therapy is not universal.

**Alga** (plural: algae) any of various primitive, chiefly aquatic, one-celled or multicellular plants that lack true stems, roots and leaves, and which usually contain the green pigment chlorophyll. Included among the algae are large kelps and other seaweeds (macroalgae) and the microscopic diatoms, dinoflagellates, golden-brown flagellates and green algae (microalgae).

**Amnesic Shellfish Poisoning** human illness caused by eating shellfish contaminated with domoic acid (e. g., from the diatom *Pseudo-nitzschia multiseries*, *P. australis*). In extreme cases this can lead to short-term memory loss (amnesia) and seizures. Also affects sea birds and marine mammals such as sea-lions.

**Antibiotic** medication used to prevent a bacterial infection from spreading, e. g., penicillin, streptomycin. Antibiotics do not work against viruses.

**Anti Microbial Resistance** the phenomenon whereby overprescription of antibiotics in human and veterinary medicine has resulted in the evolution of antibiotic resistant bacterial strains thus causing a looming global health crisis by rendering traditional drugs ineffective.

**Antiviral** medication used to prevent a viral infection from spreading.

**Archaea** a group of single-celled bacteria-like organisms (without nucleus) but genetically distinct from the real bacteria. Includes methane producers and heat tolerant deep-sea vent microbes. Not usually pathogenic to humans.

**Bacteriophages** "good" viruses that infect and destroy bacteria but are harmless to humans, plants and animals.

**Banded iron formations** layered fossil iron deposits aged two to three billion years old, marking the first appearance of oxygen in our planet's atmosphere. This – so-called Great Oxidation Event – resulted from the oxidation and subsequent precipitation of abundant iron dissolved in ancient seawater.

**Biofuels** fuels derived from recently grown biological materials such as plants, algae or organic waste. Examples are bioethanol, biodiesel and biogas. Biofuels offer a renewable alternative to fossil fuels.

**Bioluminescence** the emission of neon blue light by living organisms, e. g., by the dinoflagellate *Noctiluca*. This phenomenon results from biochemical reactions of the molecule luciferin with an enzyme called luciferase, which combined with oxygen generates light. Bioluminescence serves various functions in nature, such as for mushrooms to attract flies that spread spores, for glow worms to attract prey, and for fireflies to attract a mate. For sea creatures such as *Noctiluca*, it is thought to act as a burglar alarm to scare off predators.

 | HTTPS://DOI.ORG/10.1515/9783112228265-065

**Black Death** the most devastating human pandemic in the 14th century caused by the bacterium *Yersinia pestis* responsible for more than 50 million fatalities in Medieval Europe. The most common form (called "Bubonic Plague") results in swollen and tender lymph nodes in the groin, armpits or neck called "buboes". The rarest but deadliest form "Pneumonic Plague" affects the lungs and can be spread from person to person. This disease still is with us today and can be deadly if not treated promptly with antibiotics.

**Blue-green algae** see Cyanobacteria

**Brevetoxin** a neurotoxin produced by dinoflagellates such as *Karenia brevis* and *Karenia cristata*, which when aerosolized can lead to respiratory problems in humans from sea spray.

**Carotenoid** yellow, orange, and red pigments that are produced by plants and algae, as well as several bacteria, Archaea, and fungi. Carotenoids give the characteristic color to carrots, tomatoes, flamingos, lobster and daffodils. Examples of carotenoids are astaxanthin and β-carotene.

**Cholera** acute diarrheal illness caused by infection of the human intestine with *Vibrio cholerae* bacteria. Waterborne disease linked to poor sanitation. People can get sick when they swallow food or water contaminated with cholera bacteria. A number of safe and effective oral vaccines for cholera are available today.

**Ciguatera poisoning** human illness caused by eating certain tropical coral reef fish (e. g., coral trout, Spanish mackerel) that are contaminated with ciguatoxins (e. g., from *Gambierdiscus polynesiensis* benthic dinoflagellates).

**Coccolithophorid** small plankton algae, which are covered by delicate calcareous scales known as coccoliths, occasionally producing milky-white seas. Responsible for the geological chalk deposits (150 million years old) of the white cliffs of Dover.

**Colony** a group of the same kind of micro-organisms, which are connected and coexist in close association.

**Contagious** when a disease is spread by direct contact from person to person.

**COVID-19** contagious human disease spread by the Severe Acute Respiratory Syndrome Coronavirus 2 (SARS-CoV-2). Symptoms include respiratory difficulty, loss of taste and smell, coughing, fever, headaches and fatigue. Preventable by vaccination and treatable with antiviral drugs.

**Cyanobacteria** primitive organisms characterized by the absence of a nucleus and other membrane-bound organelles (therefore best be classified as bacteria), and their (often) blue-green color. Previously often referred as blue-green algae.

**Cyst** a thick-walled stage in the life history of algae or bacteria, also called resting spore. Cysts are often produced at the onset of adverse conditions, and when conditions are suitable, they germinate to seed new populations.

**Diarrhetic Shellfish Poisoning** nonfatal human illness caused by eating shellfish contaminated with certain dinoflagellate toxins (e. g., from *Dinophysis acuminata*)

**Diatomite (diatom ooze)** Fossil sediments containing large numbers of well-preserved diatom cell walls formed by the sinking of living diatoms from the surface water. Diatomite is used in industrial filtration such as the purification of beer, liquors, and wine, the cleansing of greases and oils, production of cement, removal of microbial contaminations, mechanical insecticide, absorbent, matting agent, fine abrasive such as in tooth paste or shoe polish, and filler in paints.

**Diatoms** single-celled golden brown colored algae, with siliceous cell walls consisting of two overlapping symmetrical parts.

**Dimethyl sulfoxide** a volatile sulfur compound produced by selected species of microalgae and seaweeds, responsible for the 'fishy' smell of the sea.

**Dinoflagellates** Single-celled golden brown algae, characteristically having a large nucleus with clearly visible (condensed) chromosomes, and with one flagellum protruding from the girdle groove and another from the sulcus groove. Some species have a cell covering of cellulose plates (armored dinoflagellates), while others are bounded by a membranous covering only (unarmored dinoflagellates). Masters of chemical warfare

**DNA (deoxyribonucleic acid)** the hereditary material found in all living organisms (and in many viruses), made up of long chains of the chemical bases adenine (A), guanine (G), cytosine (C) and thymine (T). DNA carries genetic information essential for the development and functioning of a living organism, and the transmission of inherited traits.

**Electron microscope** a microscope that uses electrons rather than light to produce magnified images, especially of objects whose dimensions are smaller than the wavelengths of visible light; see scanning electron microscope and transmission electron microscope.

**Endemic** when a certain infectious disease has continuous transmission within a given region.

**Epidemic** when more cases of a given disease than expected occur across a given region.

**Flagellum** (plural: flagella) whip-like extension of the cells of certain microorganisms, called flagellates. Mainly used for locomotion.

**Fungicide** pesticides used to kill parasitic fungi or their spores.

**Fungal toxins** (mycotoxins) fungal metabolites that affect humans, notably those with underlying health problems or a weakened immune system such as caused by chronic lung disease, tuberculosis, HIV, cancer and diabetes, or critically ill patients in hospital care. Examples: Aflatoxins, fumonisins, ochratoxins, patulin

**Fungus** (plural: fungi) a group of organisms that includes yeasts, moulds and mushrooms. They are classified in a kingdom different from plants, bacteria, and protists among others because of having chitin in their cell walls. Genetically, fungi are more closely related to animals than to plants.

**Generation time** the average time between two consecutive generations in the lineages of a population. In humans, this typically ranges from 20 to 30 years, in microalgae one to three days, in bacteria 20 to 30 minutes, and for viruses eight to 72 hours.

**Genome** the complete set of genes or genetic material present in a cell or organism. Some genomes are incredibly small, such as those found in viruses and bacteria, whereas other genomes can be unexplainably large. There does not appear to be a consistent correlation between biological complexity and genome size.

**Geosmin** the characteristic smell of earth after rain is caused by a chemical compound called geosmin produced by selected soil bacteria, cyanobacteria and some fungi.

**Green algae** single-celled algae, characterized by the presence of the olive-green pigment chlorophyll b. Precursors to the green land plants

**Human microbiome** the community of protozoa, Archaea, fungi, viruses and predominantly bacteria that live symbiotically on and within various sites of the human body, especially the gut.

**Ichthyotoxins** toxins that selectively kill fish by damaging fish gills and inhibiting their respiration. Typically fish killing toxins are not of human health concern but can cause devastating losses to the finfish aquaculture industry.

**Immunity** the ability of a biological system to resist disease, often due to the activity of specialized blood cells or the presence of antibodies produced in response to prior exposure to a given pathogen or introduced via a vaccine.

**Influenza** human disease caused by various permutations of the influenza virus. Influenza damages human lung and bronchial tissues, allowing the normal nose and throat flora to invade the lungs, causing disease and eventually death. Preventable by seasonal vaccination against ever changing mixtures of virus strains.

**Kingdoms of Life** The second highest rank of classification of life forms, below the two domains of Eukaryota (organisms with nucleus) and Prokaryota (without nucleus). To date, we recognize six kingdoms: Animalia, Plantae, Fungi, Protista, Archaea and Bacteria. Because viruses must hijack host cells before they can reproduce or synthesize proteins, they are not considered living organisms in their own right, and are not recognized as a separate kingdom. Kingdoms are further divided into the ranks of phylum, class, order, family, genus and the lowest rank of species and then strains.

**Light microscope** an optical instrument that uses a combination of glass lenses to produce magnified images of objects too small to be seen by the unaided human eye. Magnification up to 1,000×.

**Measles** human disease caused by the *Morbillivirus*. Symptoms include a high fever, cough, runny nose and a diagnostic rash all over the body. Most common in children. Vaccines became available in 1963.

**Microbe** organisms that are too small to be seen by humans without using a microscope, which in this book covers protists (algae, protozoa), microscopic fungi, bacteria, Archaea and viruses.

**Microcystin** a liver damaging toxin in contaminated drinking water, produced by cyanobacteria such as *Microcystis*

**Micron** one-thousandth of a millimeter (μm)

**Mixotrophy** the ability of an organism to combine autotrophy (as practiced by photosynthetic plants) and heterotrophy (as practiced by feeding animals). This combined strategy used by most protist plankton fundamentally has changed our understanding of the flow of energy and materials in the ocean food web, with implications for productivity, ecosystem stability, and biodiversity.

**Mycelium** a threadlike network of fungal filaments – called hyphae – mostly found in soil and around or inside plant roots

**Neurotoxins** toxins that disrupt the normal functioning of the human nervous system

**Oomycete** microscopic, eukaryotic organisms commonly referred to as water moulds. They were long mistakenly classified as fungi due to their filamentous growth. However, modern genetic analysis revealed that they belong to a group that also contains diatoms and brown algae. This means that the plant crop diseases oomycetes cause (e. g., *Phytophthora*) must be managed using methods different from those used to control fungal infections.

**Pandemic** the spread or rise of a disease over a very large geographic area

**Paradox of the Microbes** initially phrased as *Paradox of the Plankton*. How is it possible for a large number of species to coexist in a relatively unstructured environment all competing for the same resources, hence defying classical laws of competition.

**Parasite** an organism that lives in or on another different type of organism and feeds on it. Example: *Plasmodium* protozoa in mosquitoes.

**Paralytic shellfish poisoning** (PSP) human illness caused by eating shellfish that are contaminated with dinoflagellate saxitoxins (e. g., from *Alexandrium* or *Pyrodinium*). In extreme cases this can lead to death through respiratory paralysis.

**Pathogen** an organism that causes disease

**Photobioreactor** a cultivation system designed for growing microalgae or cyanobacteria using artificial light sources or sunlight to facilitate photosynthesis. These systems offer precise control over environmental factors and are widely used in industrial sectors like biofuels, nutraceuticals, cosmetics, pharmaceuticals, and wastewater treatment. By contrast fermentation vessels are used to grow bacteria, yeasts and moulds in the absence of light.

**Photosynthesis** the process by which green chlorophyll containing cyanobacteria, algae and land plants transform light energy into the chemical energy to fuel their metabolism. Also referred to as biological carbon fixation. In this process inorganic carbon (particularly carbon dioxide) is converted to organic compounds, with oxygen released as a byproduct.

**Phytoplankton** plant plankton; see Plankton.

**Plankton** plant and animal organisms, mostly microscopic, that float or drift passively with the currents (*planktos* means made to wander), both in fresh and saltwater.

**Plankton bloom** prolific growth of plankton algae; see Red tide.

**Polymerase Chain Reaction** first invented in 1983, the Polymerase Chain Reaction (PCR) is laboratory method of repeated cycles of heating and cooling to amplify copies of specific DNA sequences. This protocol has revolutionized the application of molecular genetics in forensic science, medicine and has improved our understanding of evolutionary relationships between organisms.

**Probiotics** live bacteria and yeasts that have beneficial effects on the human body, provided as dietary supplements such as yoghurt-like drinks. As opposed to *prebiotics*

which are health foods such a dietary fiber to stimulate gut microbes, while *postbiotics* are beneficial compounds produced by gut bacteria such as fermented compounds like kombucha or sauerkraut.

**Protist** an artificial grouping of mostly single-celled organisms with a true nucleus, but which are not an animal, plant or fungus. Includes microscopic algae but also seaweeds, as well as protozoa.

**Polyunsaturated fatty acids (PUFAs)** fatty acids with more than one double bond in their carbon chain, essential for human health and obtained through diet

**Quarantine** a specific period of time (traditionally "40 days") in which a person or animal that has a disease, or may have one, must stay or be kept away from others in order to prevent the spread of the disease. Quarantine practices are the cornerstone of organized responses to infectious disease outbreaks.

**Radiolarian** single-celled protists that possess delicately sculpted skeletons of silica or in the case of the subgroup *Acantharia* skeletons of strontium sulfate.

**Red tide** dense concentrations of plankton organisms (mostly dinoflagellates) that can color the sea red or brown.

**RNA (ribonucleic acid)** the sister molecule to DNA, but with different sugar (ribose instead of deoxyribose), chemical bases (uracil instead of thymine) and three-dimensional structure (single instead of double stranded). RNA participates in protein synthesis and gene regulation, but is chemically less stable than DNA. There exist various types of RNA with different functions, such as rRNA (ribosomal RNA) and mRNA (messenger RNA).

**Scanning electron microscope** a type of electron microscope in which a beam of electrons scans the outer surface of objects previously coated with a thin layer of gold or platinum. Magnification up to 30,000×

**Serotype** variation within a species of bacteria or virus, categorized by unique characteristics on its cell surface causing different binding of surface antigens. Serotypes are different from strains, which are isolates with different traits , and also different from genotypes which entails heritable genetic characteristics.

**Smallpox** a human disease caused by the *Variola* virus leaving survivors scarred with pockmarks, most prominently on the face. Today essentially eradicated by a global vaccination campaign.

**Symbiosis** a close and long-term biological interaction between two organisms of different species, both benefitting each other in some way. This process has been critical to evolution of life on our planet. Example: zooxanthella dinoflagellates in the tissues of corals

**Tintinnids** single celled protozoa, that often live in a protective flask-like house called lorica, sometimes decorated with calcareous coccolithophorid scales or diatom fragments

**Transmission electron microscope (TEM)** a type of electron microscope in which electrons pass through sections of thin biological specimens. Magnification up to 100,000×

**Tuberculosis** infectious lung disease caused by the bacterium *Mycobacterium tuberculosis* leading to inflammations of the lungs (tubercules) which may spread. Preventable with a course of antibiotics of streptomycin

**Vaccination** protection against harmful diseases that involves using the body's natural defenses to build resistance. A vaccine contains a weakened form of a given pathogen, allowing the body to create antibodies against it without succumbing to the disease.

**Vector** an animal that transmits a disease from person to person, e. g., via rats, mosquitoes and fleas

**Virus** a microscopic infectious agent that depends on a host for its own survival. Viruses contain genetic material, either in the form of RNA or DNA, but no known virus contains both.

**Yeast** bubble-like fungal cells of *Saccharomyces* species, one of the earliest domesticated microbes used in fermentation processes to make bread, beer and wine.

**Zoonosis** a disease that spreads to humans from a non-human vertebrate such as monkey, bird or bat.

**Zooxanthella** single-celled coccoid dinoflagellates that live in symbiosis with marine invertebrates including sponges, corals, jellyfish, snails, clams, flat worms but also plankton radiolarians and foraminiferans.

# Bibliography

## Introduction


1. Bynum H, Bynum W. Remarkable plants that shape our world. London: Thames & Hudson; 2014. 240 p.
2. Chaline E. Fifty animals that changed the course of history. Allen & Unwin: Australia; 2011. 223 p.
3. Haeckel E. Künstformen der Natur. 1899–1904, 144 p., Reprinted by Prestel-Verlag, New York (1998).
4. Hallegraeff, GM. Plankton: a microscopic world. Leiden: E.J. Brill Publishers; 1988. 128 p.
5. Hallegraeff GM. Plankton: a critical creation. Tasmania: Foot & Playsted Printers; 2006. 100 p.
6. McNeill W. Plagues and peoples. New York: Anchor Books; 1976. 369 p.
7. Diamond J. Guns, germs and steel. London: Random House UK; 1997. 480 p.
8. Kennedy J. Pathogenesis. London: Penguin Random House; 2023. 384 p.
9. Cavicchioli R., Ripple WJ, Timmis KN et al. Scientists' warning to humanity: microorganisms and climate change. Nature Reviews Microbiology. 2019;17:569–586.


## What are Microbes?


10. Van Leeuwenhoek A. Concerning green weeds growing in water, and some animalcula found about them. Philosophical Transactions Royal Society London. 1703;283:1304–1311.
11. Hooke R. Micrographia: or some physiological descriptions of miniature bodies made by magnifying glasses. London: Martyn and Allestry Printers; 1665. 279 p.
12. Smit H, Reijnders W, van Doorn J, van der Oost J, Willemsen P. Microcanon-Ontdekkingsreis door de microbiologie. Amsterdam: Veen Media; 2011. 246 p. (in Dutch).


## Kingdoms of Life


13. Woese CR, Kandler O, Wheelis ML. Towards a natural system of organisms: proposal for the domains Archaea, Bacteria and Eucarya. Proceedings National Academy Science USA. 1990;87:4576–4579.
14. Harris HMB, Hill C. A place for viruses on the tree of life. Frontiers Microbiology. 2021;11:604048. doi: 10.3389/fmicb.2020.604048.
15. Hug L, Baker B, Anantharaman K. et al. A new view of the tree of life. Nature Microbiology. 2016;1:16048. https://doi.org/10.1038/nmicrobiol.2016.48.


 | HTTPS://DOI.ORG/10.1515/9783112228265-066

## Lungs of the Planet: First Cyanobacteria


16. Cleeves LI, Bergin EA, Alexander CM et al. The ancient heritage of water ice in the solar system. Science. 2014;345:1590–1593.
17. Awaramik SM. The oldest record of photosynthesis. Photosynthesis Research. 1992;33:75–89.
18. Schopf JW. Cyanobacteria: Pioneers of the early Earth. Nova Hedwigia Beihefte. 1996;112:13.


## *Prochlorococcus*: Ocean's Smallest Photosynthetic Cell


19. Chisholm SW. The life and times of the ocean's smallest photosynthetic cell. In: Kolter R, Maloy S, editors. *Microbes and evolution: the world that Darwin never saw*. Washington, DC. American Society Microbiology Press; 2012. p. 165–172.
20. Tschoeke D, Salazar VW, Vidal L et al. Unlocking the genomic taxonomy of the *Prochlorococcus* collective. Microbial Ecology. 2020;80:546–558.


## *Trichodesmium* "sea sawdust"


21. Cribb AB. Sea Sawdust. Queensland Naturalist. 1969;19:115–117.
22. Devassy VP, Bhattathiri PMA, Quasim, SZ. *Trichodesmium* phenomenon. Indian Journal of Geo-Marine Sciences. 1978;7:168–186.
23. Sato S, Paranagua M, Eskinazi E. On the mechanism of red tide of *Trichodesmium* in Recife Northeastern Brazil, with some considerations of the relation to the human disease, 'Tamandare Fever'. Trabalhos Instituto Oceanografico Universidas Recife 1963/4, 5/6, 7–49.
24. Purwani J, Pratiwi E, Sipahutar IA, Husnain. The effect of different species of cyanobacteria on the rice yield and nitrogen use efficiency under different levels of nitrogen fertilizer on Alluvial West Java. *IOP Conference Series: Earth Environmental Science*. 2021;648:012196.


## Cyanobacterial Blooms and Drinking Water


25. Francis G. Poisonous Australian lake. Nature. 1878;18:11–12. doi: 10.1038/018011d0.
26. Hawkins PR, Runnegar MTC, Jackson ARB, Falconer IR. Severe hepatoxicity caused by the tropical cyanobacterium (blue-green alga) *Cylindrospermopsis raciborskii* (Woloszynska) Seenaya et Subba Raju isolated from a domestic supply reservoir. Applied and Environmental Microbiology. 1985;50:1292–1295. doi:10.1128/aem.50.5.1292-1295.

27. Humpage AR, Rositano J, Bretag AH et al. Paralytic shellfish poisons from Australian cyanobacterial blooms. Australian Journal of Marine and Freshwater Research. 1993;45:761–771. doi: 10.1071/MF9940761.
28. Jochimsen EM, Carmichael WW, An JS et al. Liver failure and death after exposure to microcystins at a hemodialysis center in Brazil. The New England Journal of Medicine. 1998;338:873–878. doi:10.1056/NEJM199803263381304.
29. Pearl H, Huisman J . Blooms like it hot. Science. 2008;320(5872):57–58.

## *Spirulina* and the Aztecs

30. Henrikson R. Spirulina Source. Richmond, USA. Available from: www.spirulinasource.com (accessed 22 June 2021).

## Microscopic Algae: Grasses of the Sea

31. Sunagawa S, Acinas SG, Bork P et al. *Tara* Oceans: towards global ocean ecosystems biology. Nature Reviews Microbiology. 2020;18:428–445. https://doi.org/10.1038/s41579-020-0364-5.
32. Hallegraeff GM. Plankton: a critical creation. Tasmania: Foot & Playsted Printers; 2006. 100 p.
33. Guiry MD. How many species of algae are there? A reprise. Four kingdoms, 14 phyla, 63 classes and still growing. Journal of Phycology. 2024. https://doi.org/10.1111/jpy.13431.
34. Andersen RA, editor. Algal culturing techniques. New York: Elsevier; 2005. 596 p.
35. Falkowski P. Ocean science: the power of plankton. Nature. 2012;483:S17–S20. https://doi.org/10.1038/483S17a.
36. United Nations Global Compact. Plankton manifesto, 19 pp. New York, USA. Available from: https://unglobalcompact.org/library/6242. 2024.
37. Grigoratou M, Menden-Deuer S, McQuatters-Gollop A et al. The immeasurable value of plankton to humanity. Bioscience. 2025. https://doi.org/10.1093/biosci/biaf049.

## Coccolithophorid *Emiliania huxleyi* and Gaia

38. Young JR, Geisen M, Cros L, Kleijne A, Probert I, Ostergaard JB. A guide to extant coccolithophore taxonomy. Journal of Nannoplankton Research. Special Issue. 2003;1:1–132.
39. Ehrenberg CG. Mikrogeologie: Das Erden und Felsen schaffende Wirken des unsichtbar kleinen selbständigen Lebens auf der Erde. Leipzig: Leopold Voss; 1854. 374 p.
40. Lovelock J. Healing Gaia: practical medicine for the planet. New York: Harmony Books; 1991. 192 p.

41. Martin JH. Glacial-Interglacial $CO_2$ change: the iron hypothesis. Paleoceanography. 1990;5(1):1–13.
42. Boyd PW, Jickells T, Law CS et al. Mesoscale iron enrichment experiments 1993–2005: synthesis and future directions. Science. 2007;315:612–617.

### *Phaeocystis* and Sea Foam

43. Peperzak L, van Wezel R. Human fatalities related to a *Phaeocystis* harmful algal bloom in the North Sea. Harmful Algae. 2023;130:102545.
44. Lancelot C, Billen G, Sournia A et al. *Phaeocystis* blooms and nutrient enrichment in the continental coastal zones of the North Sea. Ambio. 1987;16:38–46.
45. Savage RE. *Phaeocystis* and herring shoals. Journal of Ecology. 1932;20:326–340. doi:10.2307/2256082.
46. Ni J, Chen H, Dai L, Ji J, Lin J, Yu T. Early warning and monitoring of the safety risk of coastal nuclear power plant cold source under the stress from *Phaeocystis globosa* blooms. Marine Freshwater Research. 2024;75:MF23179.

### *Phytophthora infestans* and Irish Potato Famine

47. Ivanov AA, Ukladov EO, Golubeva TS. *Phytophthora infestans*: an overview of methods and attempts to combat late blight. Journal of Fungi. 2021;7(12):1071. doi: 10.3390/jof7121071.

### Diatoms: Miniature Architects

48. Round FE, Crawford RM, Mann DG. The diatoms: biology and morphology of the genera. Cambridge, UK. Cambridge University Press; 1990. 747 p.
49. Carson R. The sea around us. New York, USA. Oxford University Press, 1951.

### Blue Diatom *Haslea ostrearia*: Oysters Fit for Royalty

50. Gastineau R, Davidovich N, Hansen H et al. *Haslea ostrearia*-like diatoms: biodiversity out of the blue. In: Jacquot J-P, Gadal P, Bourgougnon N, editors. Advances in Botanical Research. Vol 71. Sea plants. New York, USA. Academic Press, Elsevier; 2014. p. 441–446.

### *Didymosphaenia geminata* or "rock snot"

51. Kilroy C, Larned S, Biggs B. The non-indigenous diatom *Didymosphenia geminata* alters benthic communities in New Zealand rivers. Freshwater Biology. 2009;54:1990–2002.
52. Kato-Unoki Y, Kurihara A, Kuge T, Shimasaki Y, Suzawa Y, Mayama S. Genetic evidence for the invasion of *Cymbella janischii* (A. Schmidt) De Toni, 1891 in Japan. BioInvasions Records. 2022;11(2):409–415.

### *Pseudo-nitzschia* and Hitchcock's *The Birds*

53. Bates SS, Hubbard KA, Lundholm N, Montresor M, Leaw CP. Pseudo-nitzschia, Nitzschia, and domoic acid: New research since 2011. Harmful Algae. 2018;79:3–43.

### Diatom Ooze and Alfred Nobel's Dynamite

54. Lizitzin AP. Distribution of *siliceous microfossils* in suspension and in bottom deposits. In: Funnel BM, Reidel WR, editors. The micropaleontology of oceans. New York. Cambridge University Press; 1971. p. 173–195.
55. Håkansson H, Hulthén B. Identification of diatoms in Neolithic pottery. Diatom Research. 1988;3:39–45. doi: 10.1080/0269249X.1988.9705015.

### *Alexandrium* and Captain Vancouver's Poison Cove

56. Vancouver G. Voyage of discovery to the North Pacific Ocean, and round the world in the years 1791–1795. London: GG and J Robinson and J Edwards; 1798. 432 p.
57. Tester PA, Matweyou J, Himelbloom B, Wright B, Kibler SR, Litaker RW. Saxitoxin and the Cold War. In: Hess P, editor. Proceedings of the 18th international conference on harmful algae, Nantes. Paris: IOC-UNESCO; 2020. p. 191–194.
58. Hallegraeff G, Bolch CJ. Unprecedented toxic algal blooms impact on Tasmanian seafood industry. Microbiology Australia. 2016;37(3):143–144.

### *Karenia brevis* and Florida Red Tides

59. Steidinger K. Historical perspective on *Karenia brevis* red tide research in the gulf of Mexico. Harmful Algae. 2009;8(4):549–561.
60. Murray SA, Bolch CJS, Brett S et al. A catastrophic marine mortality event caused by a complex algal bloom including the novel brevetoxin producer, *Karenia cristata* (Dinophyceae). Nature Ecology Evolution 2026. https://doi.org./10.1038/s41559-026-03115-0.

## *Gambierdiscus* and Ciguatera Poisoning

61. Chinain M, Gatti CM, Darius HT, Quod J-P, Tester PA. Ciguatera poisonings: a global review of occurrences and trends. Harmful Algae. 2021. 10.1016/j.hal.2020.101873.

## *Noctiluca* Night Lights

62. Hallegraeff GM, Albinsson ME, Dowdney J, Holmes AK, Mansour MP, Seger A. Prey preference, environmental tolerances and ichthyotoxicity by the red-tide dinoflagellate *Noctiluca scintillans* cultured from Tasmanian waters. Journal of Plankton Research. 2019;4:407–418.

## Harmful Algal Blooms and Finfish Kills

63. Shilo M. The toxic principles of *Prymnesium parvum*. In Carmichael, WW, editor. The water environment: algal toxins and health. New York: Plenum Press; 1982. p. 37–47.
64. GlobalHAB. Fish-killing marine algal blooms: Causative organisms, ichthyotoxic mechanisms, impacts and mitigation. In: Hallegraeff, GM et al, editors. Paris: UNESCO-IOC/SCOR; 2023. p. 96. IOC Manuals and Guides, 93. doi: http://dx.doi.org/10.25607/OBP-1964.
65. Hallegraeff GM, Anderson DM, Belin C et al. Perceived global increase in algal blooms is attributable to intensified monitoring and emerging bloom impacts. Communications Earth & Environment. 2021;2(1):1–10. doi:10.1038/s43247-021-00178-8.
66. Hallegraeff GM. Microalgae and Human Affairs: Massive Increase in Knowledge Drives Changes in Perceptions of Good and Bad Blooms In: Hendlin YH, Weggelaar J, Derossi N, Mugnai S, editors. Being algae: transformations in water, plants. Critical Plant Studies 8. Leiden: Brill; 2024. p. 175–194.

## Zooxanthellae and Coral Bleaching

67. Wild A, Reed A, Barr B, Crocetti G. Zobi and the Zoox; A story of coral bleaching. Scale Free Network. CSIRO Publishing, 2018. 28 p.
68. Van Oppen MJH, Gates RD, Blackall LL et al. Shifting paradigms in restoration of the world's coral reefs. Global Change Biology. 2017;23:3437–3448.

## *Dunaliella salina* and β-carotene

69. Borowitzka LJ, Borowitzka MA, Moulton T. Mass culture of *Dunaliella*: from laboratory to pilot plant. Hydrobiologia. 1984;116/117:115–121.
70. Borowitzka MA. Microalgae in medicine and human health: a historical perspective. In: Levine IA, Fleurence J, editors. Microalgae in Health and Disease Prevention. New York. Academic Press; 2018. p. 195–210.
71. Novoveská L, Ross ME, Stanley MS, Pradelles R, Wasiolek W, Sassi J-F. Microalgal carotenoids: a review of production, current markets, regulations, and future direction. Marine Drugs. 2019;17:640. doi: 10.3390/md17110640.
72. Wordsworth, R, Quayum, R, Kocharian, E, Pearson, A, Portillo, X, Yang, M., Cockell, CS et al. Biomaterials for organically generated habitats beyond Earth. Science Advances 2025, 11, eadp4985.

## *Botryococcus*: Good and Bad Oils

73. Kenny P, Flynn KJ. Physiology prevents commercially viable photoautotrophic production of microalgal biofuels. Journal of Applied Phycology. 2017. doi: 10.1007/s10811-017-1214-3.
74. Borowitzka MA, Moheimani NR (eds.). Algae for Biofuels and Energy. Developments in Applied Phycology 2013, 5, doi 10.1007/978-94-007-5479-9_1. Springer Science Business Media Dordrecht.
75. Zhou X-R, Yao ZJ, Benedicto K, Nichols PD, Green A, Singh S. New sustainable oil seed sources of omega-3 long-chain polyunsaturated fatty acids: a journey from the ocean to the field. Sustainability. 2023;15:11327. https://doi.org/10.3390/su151411327.

## Radiolarians: Artforms in Nature

76. Haeckel E. Künstformen der Natur, 1899–1904. Reprinted by Dover Publications, New York (1974) and Prestel-Verlag, New York (1998), 144 p.
77. Agatha S, Laval-Peuto M, Simon P. The Tintinnid Lorica. In: Dolan JR et al, editors. *The* biology and ecology of tintinnid ciliates: models for marine plankton. West Sussex, UK. John Wiley & Sons, Ltd. 2013. 17–41.
78. Hallegraeff GM, Eriksen R, Leblanc K. Agglomerated *Codonella*, *Codonellopsis* and *Dictyocysta* tintinnids from north Atlantic, tropical Pacific and Southern Ocean waters. Journal Plankton Research. 2025;47:fbaf030.

## *Plasmodium* and Malaria

79. Kwiatkowski DO. How malaria has affected the human genome and what human genetics can teach us about malaria. American Journal Human Genetics. 2005;77:171–190.
80. Gardner M, Hall N, Fung E et al. Genome sequence of the human malaria parasite *Plasmodium falciparum*. Nature 2002, 419, 498–511. https://doi.org/10.1038/nature01097.

## Baker's Yeast *Saccharomyces*

81. Parapouli M, Vasileiadis A, Afendra A-S, Hatzilouykas E. *Saccharomyces cerevisiae* and its industrial applications. AIMS Microbiology. 2020;6(1):1–31. doi: 10.3934/microbiol.2020001.

## *Penicillium*: The Birth of Antibiotics

82. Lax E. The mould in Dr. Florey's coat. London: Little Brown Publishers; 2005. 400 p.

## Soil Fungi and the Wood Wide Web

83. Wild A, Reed A, Barr B, Crocetti G. The forest in the tree. How fungi shape the Earth. Melbourne. CSIRO Publishing; 2020. 50 p.
84. Simard SW, Perry DA, Jones MD, Myrold DD, Durall DM, Molina R. Net transfer of carbon between ectomycorrhizal tree species in the field. Nature. 1997;388(6642):579–582.
85. Simard S. Finding the Mother Tree. New York: Penguin Books; 2021. 368 p.
86. Wohlleben P. The hidden life of trees. Munich, Germany: Ludwig Verlag; 2015. 272 p.
87. Albornoz FE, Dixon KW, Lambers H. Revisiting mycorrhizal dogmas: Are mycorrhizas really functioning as they are widely believed to do? Soil Ecology Letters. 2021;3:73–82. https://doi.org/10.1007/s42832-020-0070-2.

## Fungal Toxins in Food

88. Bennett JW, Klich M. Mycotoxins. Clinical Microbiology Reviews. 2003;16(3):497516. doi: 10.1128/CMR.16.3.497-516.2003.

## *Aspergillus*, Coral Disease and Dust Storms

89. Gorbushina AA, Kort R, Schulte A et al. Life in Darwin's dust: intercontinental transport and survival of microbes in the nineteenth century. Environmental Microbiology. 2007;9:2911–2922. https://doi.org/10.1111/j.1462-2920.2007.01461.x.
90. Rypien KL, Andras JP, Harvell CD. Globally panmictic population structure in the opportunistic fungal pathogen *Aspergillus sydowii*. Molecular Ecology. 2008;17:4068–4078.
91. Hallegraeff G, Coman F, Davies C et al. Australian dust storm associated with extensive *Aspergillus sydowii* fungal "bloom" in coastal waters. Applied and Environmental Microbiology. 2014;80:3315–3320.
92. Hayashi A, Crombie A, Lacey E et al. *Aspergillus sydowii* marine fungal bloom in Australian coastal waters, its metabolites and potential impact on *Symbiodinium* dinoflagellates. Marine Drugs. 2016;14(3):59. doi:10.3390/md14030059.

## Human Microbiome: You Are What You Eat

93. Buffington P, Sebghati AM, Stewart KB, Lawson S, Karaduta O. Consequences of a cesarean section on the neonatal gut microbiome and future outcomes: a qualitative analysis of the literature. Microbiota and Host 2023, 1; https://doi.org/10.1530/MAH-23-0006.
94. Daley D. The evolution of the hygiene hypothesis. Current Opinion in Allergy and Clinical Immunology 2014, 14 (5), 390–396.
95. Wild A, Hutchings B, Barr B, Crocetti G. Follow Your Gut. Scale Free Network, 2, 20.

## *Yersinia pestis* and the Black Death

96. Barbieri R, Signoli M, Chevé D et al. *Yersinia pestis*: the natural history of plague. Clinical Microbiological Reviews. 2021;34(1):e00044-19. doi: 10.1128/CMR.00044-19.

## *Vibrio* and Cholera

97. Montero DA, Vidal RM, Velasco J et al. *Vibrio cholerae*, classification, pathogenesis, immune response, and trends in vaccine development. Frontiers Medicine, Section Infectious Diseases: Pathogenesis and Therapy. 2023;10. doi.org/10.3389/fmed.2023.1155751.
98. Hassall A. Microscopic examination of the water supplied to the inhabitants of London. London: Wellcome Library; 1850. 67 p.

## *Bacillus anthracis* and Biological Warfare

99. Carlson CJ, Kracalik IT, Ross N et al. The global distribution of *Bacillus anthracis* and associated anthrax risk to humans, livestock and wildlife. Nature Microbiology. 2019;4:1337–1343. https://doi.org/10.1038/s41564-019-0435-4.

## *Myxobacterium* and Tuberculosis

100. Smith I. *Mycobacterium tuberculosis* Pathogenesis and Molecular Determinants of Virulence. Clinical Microbiology Reviews. 2003;16(3):463–496.
101. Green J. Everything is Tuberculosis. The history and persistence of our deadliest infection. New York, Random House. Crash Course Books; 2025. 208 p.

## Soil Bacteria

102. Fierer N, Wood SA, Bueno de Mesquita CP. How microbes can, and cannot, be used to assess soil health, Soil Biology and Biochemistry. 2021;153:108111. https://doi.org/10.1016/j.soilbio.2020.108111.
103. Young JPW, Moeskjær S, Afonin A et al. Defining the *Rhizobium leguminosarum* species complex. Genes. 2021;12(1):111. doi: 10.3390/genes12010111.

## Viruses

104. Zheng K, Sun J, Liang Y et al. VITAP: a high precision tool for DNA and RNA viral classification based on meta-omic data. Nature Communications. 2025;16:2226.
105. Hilleman MR. Strategies and mechanisms for host and pathogen survival in acute and persistent viral infections. Proceedings National Academy of Sciences. 2004;101: 14560–14566.

## *Variola*, Smallpox and Monkeypox

106. Senthilingam M. Outbreaks & Pandemics. London: Icon Books; 2023. 224 p.

## *Morbillivirus* and Measles

106. Senthilingam M. Outbreaks & Pandemics. London: Icon Books; 2023. 224 p.

## *Influenza* and the Spanish Flu

106. Senthilingam M. Outbreaks & Pandemics. London: Icon Books; 2023. 224 p.
107. Edler P, Schwab LSU, Aban M et al. Immune imprinting in early life shapes cross-reactivity to influenza B virus haemagglutinin. Nature Microbiology. 2024;9:2073–2083. https://doi.org/10.1038/s41564-024-01732-8.

## *Pilchard Herpes Virus* (PHV)

108. Griffin DA, Thompson PA, Bax NJ, Bradford RW, Hallegraeff GM. The 1995 Australian mass mortality of pilchard: no role found for physical or biological oceanographic factors in Australia. Australian Journal Marine Freshwater Research. 1997;48:27–42.
109. Whittington RJ, Crockford M, Jordan D, Jones B. Herpesvirus that caused epizootic mortality in 1995 and 1998 in pilchard, *Sardinops sagax neopilchardus* (Steindachner), in Australia is now endemic. Journal of Fish Diseases. 2008;31. https://doi.org/10.1111/j.1365-2761.2007.00869.x.

## *Ostreid Herpes Virus* and Pacific Oyster Mortality

110. Senthamarai MD, Rajan MR, Bharathi PV. Current risks of microbial infections in fish and their prevention methods: a review. Microbial Pathogenesis. 2023;185:106400.

## *Leporipoxvirus* and Myxomatosis

111. Kerr PJ. Myxomatosis in Australia and Europe: a model for emerging infectious diseases. Antiviral Research. 2012;93:387–415.

## Bacteriophages

112. Anonymous. Advocating for phage therapy. Nature Microbiology. 2024;9:1397–1398. https://doi.org/10.1038/s41564-024-01733-7.
113. MacLeman S et al. ATSE, CSIRO, Canberra. Available from: 2023. Curbing antimicrobial resistance. https://www.csiro.au/en/about/challenges-missions/antimicrobial-resistance/curbing-antimicrobial-resistance.

## *Human Immunodeficiency Virus* and AIDS


114. Sharp PM, Hahn BH. Origins of HIV and the AIDS pandemic. Cold Spring Harbour Perspectives in Medicine. 2011. doi 10.1101/cshperspect.a006841.
115. Bryson, B. A short history of nearly everything. Sydney: Random House; 2005. 624 p.


## *Coronaviruses* and COVID-19


116. MacIntyre R. Dark Winter. Sydney, Australia: New South Publishing; 2022. 249 p.
117. Chen X, Kalyar F, Chughtai AA, MacIntyre CR. Use of a risk assessment tool to determine the origin of severe acute respiratory syndrome coronavirus 2 (SARS-CoV-2). Risk Analysis. 2024;1–11. https://doi.org/10.1111/risa.14291.
118. Gates B. How to prevent the next pandemic. UK: Penguin Random House; 2022. 292 p., 2022.
119. Fonseca T, Edo C, Vilke JM et al. Impact of face masks weathering on the mussels *Mytilus galloprovincialis*. Water Emerging Contaminant Nanoplastics. 2024;3:3. https://dx.doi.org/10.20517/wecn.2023.57.


## The Microbial World Darwin Never Knew


120. Kolter R, Maloy S (eds). Microbes and evolution: the world that Darwin never saw. Washington DC: American Society for Microbiology Press; 2012. 299 p.
121. Hutchinson GE. The paradox of the plankton. American Naturalist. 1961;95:137–145.
122. Behrenfeld MJ, Bisson KM, Boss E, Gaube P, Karp-Boss L. Phytoplankton community structuring in the absence of resource based competitive exclusion. PLoS ONE. 2022;17(9):e0274183. https://doi.org/10.1371/journal.pone.0274183.
123. Flynn KJ, Mitra A, Wilson WH et al. 'Boom-and-busted' dynamics of phytoplankton–virus interactions explain the paradox of the plankton. New Phytologist. 2022;234(3):990–1002.


## Microevolution: The Power of Growing Fast


124. Wang RJ, Al-Saffar SI, Rogers J, Hahn MW. Human generation times across the past 250,000 years. Science Advances. 2023;9:eabm7047. doi: 10.1126/sciadv.abm7047.
125. Kwiatkowski DO. How malaria has affected the human genome and what human genetics can teach us about malaria. American Journal Human Genetics. 2005;77:171–190.

## Microbes and their Toxins

126. De Wit, R, Bouvier T. 'Everything is everywhere, but, the environment selects'; what did Baas Becking and Beijerinck really say? Environmental Microbiology. 2006;8:755–758.

## Quarantine for Ships, Aquaculture and Agriculture

127. Hallegraeff GM, Bolch CJ. Transport of diatom and dinoflagellate resting spores in ships' ballast water: Implications for plankton biogeography and aquaculture. Journal of Plankton Research. 1992;4:1067–1084.
128. International Maritime Organization, International convention for the control and management of ships' ballast water and sediments. Available from: http://www.imo.org. London, 2004.
129. Salem H, Kaltenpoth M. The Nagoya Protocol and its implications for microbiology. Nature Microbiology. 2023;8:2234–2237. doi.org/10.1038/s41564-023-01532-6.

## Quarantine for Human Disease Outbreaks

130. Rodríguez-Frías F, Quer J, Tabernero Det al. Microorganisms as shapers of human civilization, from pandemics to even our genomes: villains or friends? A historical approach. Microorganisms. 2021;9(12):2518. https://doi.org/10.3390/microorganisms 9122518.
131. Tognotti E. Lessons from the history of quarantine, from plague to influenza A. Emerging Infectious Diseases. 2013;(2):254–259. doi: 10.3201/eid1902.120312.
132. Diani S, Leonardi E, Cavezzi A et al. SARS-CoV-2-The role of natural immunity: A narrative review. Journal Clinical Medicine. 2022;11:6272. doi: 10.3390/jcm11216272.
133. Blake JS, Gilman N. Children of a modest star. California, US: Stanford University Press; 2024. 326 p.

## Microbes and Climate Change

134. Timmis K, Cavicchioli R, Garcia JL et al. The urgent need for microbiology literacy in society. Environmental Microbiology. 2019;21:1513–1528. https://doi.org/10.1111/1462-2920.14611.
135. Karačić A, Hammer P, Liberati-Pršo A-M, Krumbiegel C, Moskalenko O. The impact of dog ownership on the gut microbiota. Microbiota and Host. 2025;3. https://doi.org/10.1530/MAH-25-0006.

# Index of Subject

## (People's Names in bold)


**Abu Bakr al-Razi** 123
Acantharia 75, 164
Acquired Immune Deficiency Syndrome (AIDS) 139
aerosols 57
Aflatoxins 93
**Alexander Fleming** 2, 89
**Alexandre Yersin** 105, 152
**Alfred Hitchcock** 49
**Alfred Hoffmann** 93
**Alfred Nobel** 51, 171
algae IX, XI, 5, 9, 17, 25, 29, 30, 33, 37, 41, 53, 55, 64, 65, 67, 81, 160, 169, 172
allergy 104, 175
Amnesic Shellfish Poisoning 49
animalcules IX, 5
antibiotics 89, 159
anti-malarial drugs 81
AntiMicrobial Resistance 89
Antiretroviral (ARV) drugs 139, 159
**Anton Van Leeuwenhoek** 5
Archaebacteria 1, 9, 13
**Aristotle** 39, 61
artemisinin 81, 152
Artforms in Nature 1, 173
aspergillosis 97
asthma 104
atmospheric nitrogen 19, 115
Avatar 85
Aztecs 25, 169

**Baas-Becking** 150
*Bacillus* Calmette-Guérin vaccin 111
Bacteria 9, 101, 103, 115, 147, 167
ballast water 151, 178, 179
banded iron formations 13
**Barry Marshall** 103
beer 51, 79, 85, 87, 160, 165
Bering Sea 49
biodiesel 71
biofuels 25, 173
bioluminescence 53, 61
biotechnology 25, 69, 89
Black Death 2, 105, 152, 153, 175
blue-green algae 13
**Bob Andersen** 30
Book of Exodus 2
Botox 3, 149
**Botticelli** 37
bread 87, 165
brevetoxins 57
Bubonic Plague 105
**Buckminster Fuller** 75
burglar alarm 61, 159

Caesarean birth delivery 103
Camembert 89
**Camus** 111
**Captain Cook** 19, 59
**Carl Linnaeus** 150
**Carl Woese** 1, 9
carotenoid 69
**Charles Darwin** 29, 65, 69, 97, 145
chlorophyll *b* 17, 67
chloroplasts 13, 67, 81
chloroquine 81
Cholera 107, 152
**Chopin** 111
**Christopher Andrewes** 125
Ciguatera Poisoning 53, 59, 172
ciguatoxins 59, 149
cliffs of Dover 33
climate change 2, 49, 153, 155
Coccolithophorids 33
coccoliths 33
coorongite 71
coral bleaching 59, 65, 97, 172
coral fan 97
Coral Sea 19
Coronaviruses 141, 178
COVID-19 141, 151, 152, 178
**Craig Venter** 29
Cretaceous 33
cyanobacteria 1, 9, 13, 17, 19, 21, 25, 67, 115, 162, 168
cylindrospermopsins 21, 149

DDT 81
diatom ooze 51
diatomite 51
diatoms IX, 41, 45, 47, 51, 67, 149, 160, 171
Didymo 47


 | HTTPS://DOI.ORG/10.1515/9783112228265-067

Dihé 25
dimethyl sulphoxide 33, 37
dinoflagellates 53, 59, 67, 175
dome of St Sophia 51
domoic acid 49, 149, 171
drinking water 1, 21, 22, 103, 168
dust storms 97

**Edward Jenner** 2, 121, 152
**Ernest Haeckel** 1, 9, 75
**Ernst Chain** 89
**Ernst Ruska** 6
Eubacteria 1, 9
eukaryotic 13, 67
**Evelyn Hutchinson** 145
Exodus 53

face masks 3, 113, 141, 153, 157, 178
flamingos 69
**Florence Nightingale** 111
Foot and mouth disease 133
**Fred Hoyle** 13
fumonisins 93
fungicides 39, 95

Gaia hypothesis 33
Gambier Islands 59
**Gary Powers** 55
genetically modified crops 71
genome 2, 17, 87, 174, 178
geodesic dome 75
**George Vancouver** 55
geosmin 115, 162
Giardiasis 22
**Gottfried Ehrenberg** 33
Gram-negative bacteria 89, 101
Gram-positive bacteria 89, 101
grapevine 95
Great Oxidation Event 13
Great Plague of London 105
gut-brain axis 103

Haber-Bosch process 19
**Hans Christian Gram** 101
Harmful Algal Blooms 55, 63
hemoglobin 67
**Howard Florey** 2, 89, 151
Human Genome Project 29, 154
human microbiome 103
hydrocarbons 71
hygiene revolution 157

ichthyotoxins 63
Influenza 2, 125, 126, 152, 153, 162, 177
insulin 87
International Convention on the Control and Management of Ship's Ballast Water and Sediments 151
International Maritime Organisation (IMO) 151
Irish potato famine 39, 170
iron fertilisation 34

**James Cameron** 85
**James Cook** 59
**James Lovelock** 33
**Jared Diamond** 2
**Jean-Luc Mouget** 45
**Joan Miro** 75
**John Snow** 107, 152
**Jonathan Kennedy** 2

**Kafka** 111
**Karen Steidinger** 57
**King Tutankhamun** 94
Kingdom Fungi 1
Künstformen der Natur 75, 167, 173

Lake Alexandrina 21
Lake Erie 21
Legionnaire's disease 113
legumes 19, 115
**Louis Pasteur** 87, 89, 111
LSD 93
luciferin 61, 159
**Lynn Margulis** 67

maitotoxins 60
Malaria 81, 147, 152
marennine 45
Mars 33, 69
**Martinus Beyerinck** 1, 119
**Max Knoll** 6
Measles 2, 123, 152, 153, 176
MERS (Middle East Respiratory Syndrome) 141
**Michel Felix Dunal** 69
microbial literacy 155
microcephaly 127
microcystins 21, 149, 169
Micropia 5
mildew 95
mitochondria 13, 67
mixotrophy 53
Monkeypox 121

Montreal 1967 Expo 75
mosquitoes 2, 81, 127, 135, 153, 157
mRNA vaccines 3, 142
mycelium 85, 91, 163
mycorrhizal network 91
mycotoxins 93
Myxomatosis 135, 177

Nagoya protocol 151
**Napoleon** 111
NASA 25, 33
Neolithic pottery 51, 171
Neurotoxic Shellfish Poisoning 57
New Zealand 21, 47, 131, 171
nitrogen-fixing bacteria 115
nodularin 21
Novo Nordisk 87

ocean acidification 59
Ochratoxin 93
oil 14, 71
oomycete 39, 95
oysters 1, 45, 131, 170
ozone 14

Pacific Oyster Mortality Syndrome (POMS) 131
Palm Island 21
Paradox of the Plankton 145
Paralytic Shellfish Poisoning 53, 55
**Pasteur** 87, 89, 105, 121
**Patrick Laidlaw** 125
Patulin 93
**Paul Karrer** 69
**Paul Klee** 75
penicillin 2, 85, 89, 137, 149
phages 137
**Pharaoh Ramses V** 121
photosynthesis 13, 33, 41, 168
phytoplankton 29, 178
Plague of Justinian 105
Platypus Bay 59
pneumonia 126, 139, 141
Pneumonic Plague 105, 160
Poison Cove 55, 171
polymerase chain reaction 9, 163
Porte Monumentale 75
prebiotics 104
probiotics 103
Protista 1, 9, 75
PUFAs 71, 164
Pulmonary Tuberculosis 111
purple sulphur bacteria 67

quarantine 2, 47, 105, 133, 150, 151, 152, 153, 179
Quinine 81

rabbits 135
**Rachel Carson** 42
Radiolaria 75
Rapid Antigen Test (RAT) 55, 142
reactive oxygen radicals 63
red blood cells 5, 67, 81
Red Sea 19
red tides 19, 53, 171
**René Binet** 75
RNA viruses 119, 147
**Robert Hooke** 5
**Robert Koch** 107, 111, 152
**Robert Whittaker** 1, 9
**Robin Warren** 103
rock snot 47, 171
**Ronald Ross** 81, 152
Royal Society 5, 41, 167
**Rudolph Virchow** 139

**Saddam Hussein** 109
Sahara dust 97
Saint Anthony's fire 93
**Sallie Chisholm** 17
salmon aquaculture 63, 69
saprophytic 85
SARS (Severe Acute Respiratory Syndrome) 141
saxitoxin 55, 149
saxitoxins 3, 21, 149
Scanning Electron Microscope 6
sea foam 37, 170
serial endosymbiosis 67
shingles 122
silica wall 41
slime mould 39
sloth fever 127
smallpox 1, 2, 101, 119, 121, 123, 153, 176
social media 153
Southern Ocean 37, 42, 173
soy sauce 97
Spanish flu 125
**Stravinsky** 111
streptomycin 85, 89, 111, 149
submarine 61
**Suzanne Simard** 91

Swedish Academy of Sciences 51
Sydney IX, 22, 59, 129, 131

Tamandare fever 19
Tara Oceans 29, 169
The Birds 49, 149
**Theodor Schwann** 87
The Origin of Species 29, 145
**Thomas Huxley** 33
tintinnids 79, 173
Toledo 21
Transmission Electron Microscope 6
Trichothecenes 93
Tuberculosis 111, 152

United Nations Biological Weapons
Convention 109
United States Food and Drug Administration 55
US space station 51

vaccination passports 154, 157
vaccines 2, 121, 126, 141, 152, 157
**Van Leeuwenhoek** IX, 5, 167
variolation 121
**Vasily Kandinsky** 75
Viruses 1, 9, 119, 147, 176

**William McNeill** 2
**Wilson Smith** 125
wine 51, 79, 87, 89, 93, 95, 160
wood rot 91
wood wide web 91
World Health Organisation 107, 121, 127, 154

yeast 1, 87, 174
Yellow fever 81
Yellow Sea 19

Zearalenone 93
zero quarantine risk 154
zoonosis 139
zooxanthella 65

# Index of Species


*Agrobacterium* 39, 115
*Agrobacterium radiobacter* 39
*Agrobacterium tumefaciens* 39
*Alexandrium catenella* 53, 55
*Alternaria helianthi* 115
*Anthrax* 3, 89, 109
*Apthovirus* 133
*Arthrospira* 25
*Aspergillus flavus* 93
*Aspergillus niger* 89
*Aspergillus oryzae* 97
*Aspergillus parasiticus* 93
*Aspergillus sojae* 97
*Aspergillus sydowii* 97, 175
*Azospirillum* 115
*Azotobacter* 115

*Bacillaria* 41
*Bacillus anthracis* 109
*Bacillus megaterium* 115
*Bacillus subtilis* 115
*Bifidobacterium* 103
*Botryococcus* 71
*Botrytis cinerea* 95
*Byssochlamys* 93

*Campylobacter* 103
*Canola* 71
*Chaetomium globosum* 91
*Chattonella marina* 63
*Chlorella* 67, 71
*Chlorogloea* 19
*Clathrocanium reginae* 75
*Claviceps purpurea* 93
*Clostridium botulinum* 3, 149
*Clostridium difficile* 103
*Clostridium tetani* 149
*Coprococcus* 104
*Cryptococcus* 126
*Cryptosporidium* 22
*Cylindrospermopsis raciborskii* 21, 168
*Cymbella janischii* 47, 171

*Dialister* 104
*Didymosphaenia geminata* 47, 170
*Dolichospermum circinale* 21
*Dunaliella salina* 69

*Emiliania huxleyi* 33, 169
*Entophysalis granulosa* 13
*Eoentophysalis belcherensis* 13
*Escherichia coli* 103, 150, 151
*Eurycolium* 17

*Faecalibacterium prausnitzii* 104
*Flavobacterium* 115
*Fusarium* 93

*Gambierdiscus toxicus* 53, 59
*Giardia* 22
*Glomus* 91
*Gluconobacter* 115
*Gymnodinium catenatum* 53

*Haematococcus influenza* 126
*Haematococcus pluvialis* 69
*Haslea ostrearia* 45, 170
*Helicobacter pylori* 104
*Herbaspirillum* 115

*Karenia brevis* 53, 57, 171

*Lactobacillus* 103
*Legionella pneumophila* 113
*Leporipoxvirus* 135, 177

*Methanobrevibacter smithii* 9
*Microcystis aeruginosa* 21
*Microcystis panniformis* 21
*Moraxella catarrhalis* 126
*Morbillivirus* 123, 152, 162, 176
*Mycobacterium bovis* 111
*Mycobacterium tuberculosis* 111, 165, 176
*Mycoplasma pneumoniae* 126

*Nannochloropsis* 71
*Noctiluca scintillans* 61, 172
*Nodularia spumigena* 21
*Nostoc* 19

*Oropouche Virus (OROV)* 127
*Ostreid Herpes Virus (OsHV-1)* 131

*Paramecium bursarium* 67
*Penicillium camembertii* 89


 | HTTPS://DOI.ORG/10.1515/9783112228265-068

*Penicillium chrysogenum* 89
*Penicillium notatum* 89
*Penicillium roqueforti* 89
*Phaeocystis* 37, 170
*Phormidium autumnale* 21
*Phytium* 39
*Phytophthora infestans* 39, 170
*Phytophthora* 39, 151
*Phytophthora cinnamomi* 39
*Pilchard Herpes Virus PHV* 129
*Plasmodium falciparum* 81, 152, 174
*Plasmopora viticola* 95
*Pneumocystis* 126
*Prochlorococcus* 17
*Prolificoccus* 17
*Prymnesium parvum* 63, 172
*Prymnesium polylepis* 63
*Pseudoanabaena* 19
*Pseudochattonella verruculosa* 63
*Pseudomonas fluorescens* 115
*Pseudo-nitzschia australis* 49
*Pseudo-nitzschia multiseries* 49
*Puccinia graminis* 39
*Pyrodinium bahamense* 53, 55
*Pyxidicula* 75

*Rabbit Calicivirus* 135
*Rhizobium* 19, 115, 176
*Rhizoctonia solani* 115
*Riococcus* 17
*Roseburia intestinalis* 104
*Ruminococcus gnavus* 104

*Saccharomyces cerevisiae* 87, 165, 174
*Saccharomyces ellipsoideus* 87
*Salmonella* 103
*Sanguinea nivaloides* 69
*Shigella* 103
*Spirulina* 25, 169
*Staphylococcus aureus* 89
*Streptococcus aureus* 126
*Streptococcus pneumoniae* 126
*Streptomyces griseus* 89
*Symbiodinium* 65, 97, 175

*Thaumococcus* 17
*Tobacco Mosaic Virus* 1, 119
*Trichodesmium* 19, 168

*Varicella* 122
*Variola major* 121
*Variola minor* 121
*Vibrio* 103, 107, 149, 151, 152, 160, 175
*Vibrio cholerae* 107, 151, 152, 160, 175

*Yersinia* 2, 103, 105, 152, 175
*Yersinia pestis* 2, 105, 152, 175

Zika *Flavivirus* 127

www.ingramcontent.com/pod-product-compliance
Lightning Source LLC
LaVergne TN
LVHW081915220926
848675LV00028B/316
* 9 7 8 3 1 1 9 1 4 2 3 3 5 *